GEL ELECTROPHORESIS:
NUCLEIC ACIDS

The INTRODUCTION TO BIOTECHNIQUES series

Editors:

J.M. Graham Merseyside Innovation Centre, 131 Mount Pleasant, Liverpool L3 5TF

D. Billington School of Biomolecular Sciences, Liverpool John Moores University, Byrom Street, Liverpool L3 3AF

Series adviser:

P.M. Gilmartin Centre for Plant Biochemistry and Biotechnology, University of Leeds, Leeds LS2 9JT

CENTRIFUGATION
RADIOISOTOPES
LIGHT MICROSCOPY
ANIMAL CELL CULTURE
GEL ELECTROPHORESIS: PROTEINS
PCR
MICROBIAL CULTURE
ANTIBODY TECHNOLOGY
GENE TECHNOLOGY
LIPID ANALYSIS
GEL ELECTROPHORESIS: NUCLEIC ACIDS

Forthcoming titles

PLANT CELL CULTURE
LIGHT SPECTROSCOPY
MEMBRANE ANALYSIS

GEL ELECTROPHORESIS: NUCLEIC ACIDS

Robin Martin

Group Leader Wound Healing Research, Blond McIndoe Centre, Queen Victoria Hospital, East Grinstead, West Sussex RH19 3DZ, UK

Formerly at:

Department of Molecular Biology and Biotechnology, The University of Sheffield, Sheffield SI0 2UH, UK

Routledge
Taylor & Francis Group

LONDON AND NEW YORK

Taylor & Francis Publishers Limited, 1996

First published 1996

Routledge is an imprint of the Taylor & Francis Group, an informa business

Copyright © 1996 Taylor & Francis

A CIP catalogue record for this book is available from the British Library.

ISBN 1 872748 28 7

Taylor & Francis Publishers Ltd
2 Park Square, Milton Park, Abingdon, Oxon, OX14 4RN
270 Madison Ave, New York NY 10016

Transferred to Digital Printing 2009

DISTRIBUTORS

Australia and New Zealand
  DA Information Services
  648 Whitehorse Road, Mitcham
  Victoria 3132

India
  Viva Books Private Limited
  4325/3 Ansari Road
  Daryaganj
  New Delhi 110002

Singapore and South East Asia
  Toppan Company (S) PTE Ltd
  38 Liu Fang Road, Jurong
  Singapore 2262

USA and Canada
  Books International Inc.
  PO Box 605, Herndon, VA 22070

Typeset by Chandos Electronic Publishing, Stanton Harcourt, UK.

Publisher's Note
The publisher has gone to great lengths to ensure the quality of this reprint but points out that some imperfections in the original may be apparent.

Contents

PART 2: TECHNIQUES AND APPLCATIONS

Appendices

Abbreviations

ADPL	allele discrimination by primer length
AP	alkaline phosphatase
ASO	allele-specific oligonucleotide
BAC	N,N'-bis-acrylylcystamine
bp	base pairs
CAT	chloramphenicol acetyltransferase
CHEF	contour homogeneous electric field
dATP	deoxyadenosine triphosphate
dCTP	deoxycytosine triphosphate
dd	dideoxy
DEAE	diethylaminoethyl
dGTP	deoxyguanine triphosphate
dITP	deoxyinosine triphosphate
DNase	deoxyribonuclease
dNTP	deoxynucleotide triphosphate
DSCP	double-strand conformational polymorphism
dTTP	deoxythymine triphosphate
EDTA	ethylenediaminetetraacetic acid
EEO	electroendosmosis
EMBL	European Molecular Biology Laboratory
EMSA	electrophoretic mobility shift assay
GAA	acid α-glucosidase
GEMSA	gel electrophoretic mobility shift assay
HRP	horseradish peroxidase
Kb	kilobases
LDLr	low-density lipoprotein receptor
Mb	megabases
MDE	mutation detection enhancement
PCR	polymerase chain reaction
PEA1	plastid envelope ATPase
PTH	parathyroid hormone
RFLP	restriction fragment length polymorphism
RNase	ribonuclease
RT–PCR	reverse transcription–polymerase chain reaction
SDS	sodium dodecyl sulfate
SSC	sodium chloride, sodium citrate buffer
SSCP	single-strand conformational polymorphism
TAE	Tris-acetate-EDTA
TBE	Tris-borate-EDTA
TEMED	N,N,N',N'-tetramethylethylenediamine
TOTO-1	benzothiazalium-4-quinolinolium dimer
XP	xeroderma pigmentosum
YOYO-1	benzoxazolium-4-quinolinolium dimer

Preface

Simply stated there are two groups of scientists with an interest in nucleic acid gel electrophoresis. There are those specialists who are developing new electrophoretic techniques and are investigating the physical processes which take place during electrophoretic migration. Then there are those scientists who see the electrophoresis of DNA and RNA as a technique to be used as part of their investigations. Both groups are essential to maintain the forward progress of molecular biological knowledge, but the focus of this book is predominantly from the perspective of the second sort of scientist. The purpose of the *Introduction to Biotechniques* series is to provide an introductory review of methods which are fundamental to the biological sciences. The gel electrophoresis of nucleic acids undoubtedly qualifies in this respect. Molecular biology would scarcely exist without some form of gel electrophoresis for separating DNA and RNA molecules during their analysis or isolation.

This book does not provide step by step recipes for carrying out electrophoretic procedures. That has been done before, largely for more experienced investigators, and sources for obtaining this information are indicated in this book. What does not seem to have been available is a broader view for the beginner, which connects the nuts and bolts of the technique to the application of gel electrophoresis, within the setting of real biological investigations. In this volume, the theory and practical necessities of electrophoretic techniques are covered in the first part of the book. For the use of nondenaturing agarose gels, which is perhaps the most commonly used technique, I have gone into some detail to show how altering various parameters affects the resolution and detection which can be obtained. The same principles can be applied to techniques involving polyacrylamide gels or denaturing conditions to refine those applications. In the second part, the application of the different techniques are reviewed by looking at published research articles. In this way nucleic acid electrophoresis is put into the context of the projects for which it is an indispensable tool. This approach seems to offer the greatest benefit to the proposed readers of this book: undergraduate and postgraduate students and researchers in all biological and medical disciplines who wish to avail themselves of the power of modern molecular biology.

Robin Martin

Acknowledgements

I would like to thank collectively all the authors and publishers who gave permission to reproduce material. They are acknowledged individually in the text. Special thanks are due to Tom Armour and colleagues in the Physics and Astronomy departments at the University of Sussex, for their efforts in securing a photograph for Figure 2.7.

1 Introduction: the Variety and Forms of Nucleic Acids

1.1 Overview

The purpose of this book is to provide a newcomer to the field of molecular biology with a helpful introduction to the analysis of nucleic acids by gel electrophoresis. To qualify as a newcomer you may be an undergraduate or a postgraduate student. Alternatively, there are a great many 'new' research workers from outside molecular biology who also need to understand the principles of nucleic acid electrophoresis. Molecular biology is now so pervasive an approach that scientists and clinicians with many years of experience in their own disciplines often find themselves embarking on a molecular biology project with only a sketchy understanding of the techniques they propose to employ. The aim of this book is not to teach you to become specialists in electrophoretic technology. Rather, the intention is to give a clear explanation of the basic concepts of nucleic acid electrophoresis so that you will be able to exploit these techniques to their full potential. An appreciation of the principles of nucleic acid electrophoresis will also be of considerable benefit in getting to grips with some real research papers. Because of the pressure of space, articles in molecular biology journals usually take for granted knowledge of electrophoretic techniques. If you do not have these techniques at your fingertips, research papers can seem more complicated than in reality they are. Throughout the book those molecular biological terms that may be unfamiliar to some readers have been included in a Glossary.

1.2 Electrophoresis and the properties of nucleic acids

The word 'electrophoresis' comes from the Greek *electron* and the Latin *phore*. *Electron* is the Greek word for amber, a substance that can be charged to a high voltage of static electricity by simply rubbing with a cloth. *Phore* is the Latin word for bearer. The process of electrophoresis refers then to the electrical charges 'carried' by the molecules. The use of the term dates from 1909 when it was used to describe the movement of colloidal particles in an electric field [1]. To help visualize the concepts which I am going to discuss, the essential features of nucleic acid electrophoresis are illustrated in *Figure 1.1*. Nucleic acids are negatively charged molecules. Under the influence of an electric field they migrate towards the positive electrode. The greater the voltage, the faster they move. The medium they move through and their overall shape both affect their progress. It follows that different sizes and forms of nucleic acid move at different rates, and this provides a basis for their separation. Separating mixtures of nucleic acids as a prelude to their study, or equally importantly their purification and recovery, lies at the core of many of the techniques in molecular biology.

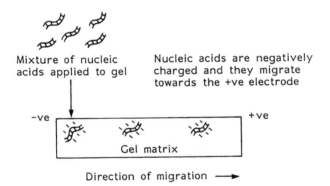

FIGURE 1.1:   *The essential features of nucleic acid gel electrophoresis. Nucleic acids are negatively charged molecules. When a mixture of different nucleic acids is allowed to move through a gel matrix they can be separated on the basis of their shape or size. Gel electrophoresis is the most important molecular biological technique for both the analysis and the purification of nucleic acids.*

There are a number of different methods that come under the umbrella of nucleic acid electrophoresis. By and large, each of these methods do not represent alternative ways of achieving the same outcome. Instead, each form of nucleic acid electrophoresis is suited to a particular task. The technique you use depends on what you wish to achieve. In general, the electrophoretic method is most often determined by the resolution required.

A nucleic acid can be considered as a molecule that is composed of three different molecular substances: the base, the sugar and the phosphate (*Figure 1.2*). It is the nature of these components that give nucleic acids their properties and it is these which are exploited in gel electrophoresis. This is not the place for a lengthy discussion of nucleic acid structure. Further details of this and the transmission and expression of nucleic acids can be found in textbooks of molecular and cellular biology [2–5]. It is important though to remember two aspects of the properties of nucleic acids that are fundamental to their behavior during electrophoresis. Firstly, nucleic acids are acids because the phosphate groups give up their H^+ ions very readily. DNA and RNA are therefore negatively charged in most buffer systems. (A buffer with a very low pH, in other words a high concentration of H^+ would be required to cause association of the H^+ ions and their phosphate groups.) If you remember this, you will never suffer the embarrassment of connecting up a gel electrophoresis apparatus the wrong way round. Nucleic acids are negatively charged and they migrate towards the positive electrode (*Figure 1.1*). The second aspect of nucleic acid structure that is of importance for nucleic acid electrophoresis concerns the inherent base pairing properties of nucleic acids. Complementary, or homologous, sequences will try to pair with each other: A opposite T (or U in RNA) and G opposite C. The greater the uninterrupted stretch of homology, the more stable the structure. Thus, in ordinary conditions, double-stranded molecules will stay double-stranded. Single-stranded molecules will form intermolecular or intramolecular duplexes (base-paired regions) wherever such opportunities exist, unless conditions are adjusted to prevent this. Intramolecular base pairing is often referred to as secondary structure. In contrast, primary structure refers to the sequence of bases in a single strand. Primary structure is formed with covalent bonds whilst secondary structure is exclusively noncovalent in nature. With these two facts in mind, there follows a review of the types of nucleic acid molecules found in nature or created in the laboratory during the course of analysis.

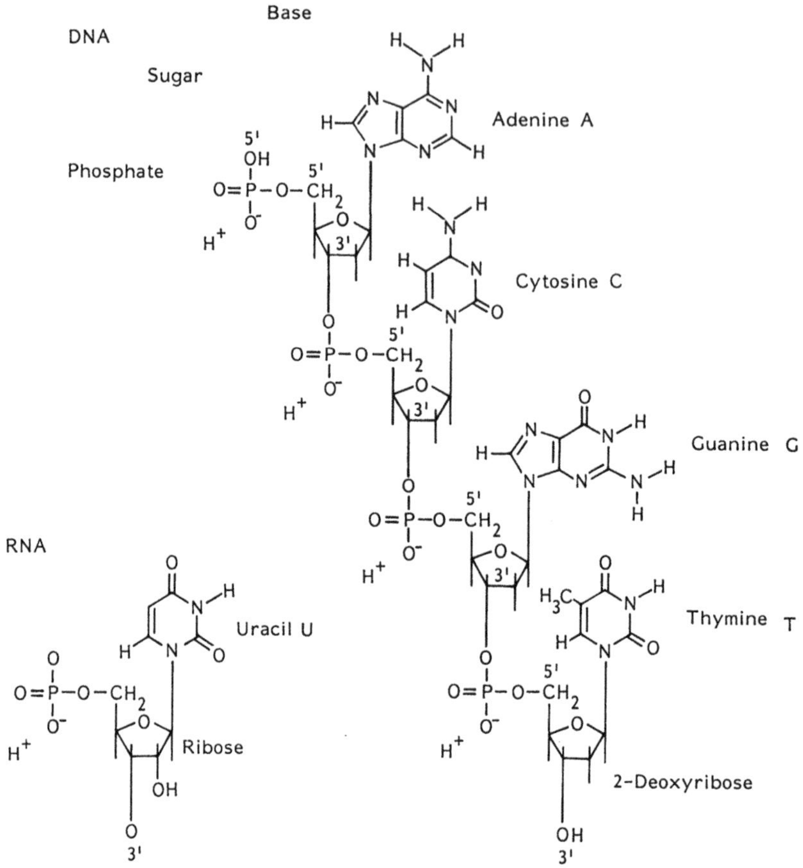

FIGURE 1.2:   *The three components of nucleic acids. Nucleic acids are constructed from three components: the bases adenine, cytosine, guanine and thymine in DNA (uracil replaces thymine in RNA); the sugar 2-deoxyribose in DNA (ribose replaces 2-deoxyribose in RNA); and the phosphate. The backbone of the molecule is formed from the alternating phosphate–sugar–phosphate–sugar chain. The phosphate groups ionize readily and release H+. DNA and RNA molecules are thus acidic and carry a negative charge. Pairing occurs between the bases A and T (U in RNA), and between G and C. The section of nucleic acid shown could pair with an identical molecule if it was rotated through 180°. Redrawn from ref. 2, Molecular Biology of the Gene, Vol. I, Fourth Edition, by Watson et al. Copyright (©) 1987 by James D. Watson. Published by The Benjamin/Cummings Publishing Company.*

1.3 The variety and forms of nucleic acid

Nucleic acids are often very long polymers. An mRNA molecule that contains 5000 bases is not unusual and is certainly not the record. DNA molecules can be very long indeed. The human genome contains 3000 million bases. With 23 chromosomes, it is clear that there are double-stranded linear DNA molecules in each of our body cells that contain hundreds of millions of bases. One base pair spans 0.34 x 10^{-9}m. Individual chromosomes can therefore extend to several centimeters. All the DNA in a single human cell would stretch for 2 meters in uncondensed form. Molecules of this size cannot be analyzed by electrophoresis without fragmentation into smaller sizes. There are essentially two methods that can do this:

(1) cutting DNA at defined sites with restriction enzymes;
(2) amplifying defined sections of DNA, or DNA copies of RNA, using the polymerase chain reaction (PCR).

Understanding these two techniques is as crucial for an appreciation of molecular biology as understanding the technique of nucleic acid gel electrophoresis. There are a number of excellent books that cover these two areas at the same level as the present text [6,7].

Figure 1.3 shows most of the different kinds of nucleic acids that are commonly encountered. Starting with DNA, the various permutations

Nucleic acid

DNA

1 Linear double stranded DNA molecules

• Complete viral genomes e.g. *Herpes*
• Whole eukaryotic chromosomes
• Randomly sheared or digested DNA
• DNA digested with restriction enzymes
• Products of Polymerase Chain Reactions (PCR)

FIGURE 1.3: *The variety and forms of nucleic acids. A number of different forms of nucleic acids are drawn in this figure using a 'ladder' format, where the rungs of the ladder indicate base pairing and the phosphate–sugar backbone is represented by the sides of the ladder. In each case, examples of where these particular forms of nucleic acids might be encountered in nature or in the laboratory are indicated. In Chapter 6 each of the forms described here are matched to particular techniques of gel electrophoresis.*

2   Supercoiled, double stranded circular DNA molecules

- Plasmids e.g. pBR322, pUC19
- Viral genomes e.g. *Papilloma virus, SV40 virus*
- Circular molecules are often present in cells as a supercoiled form. This is caused by breaking the molecule, twisting one end several times and sealing the joint. The enzymes which accomplish this are known as topoisomerases and gyrases

3   Nicked double stranded circular DNA molecules

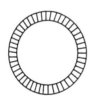

- When single strand nicks are present, supercoiling is removed (relaxed)

4   Covalently-closed, double stranded circular DNA molecules

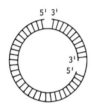

- Covalently-closed circular DNA molecules are formed when single stranded nicks in super-coiled molecules are sealed with DNA ligase

- Covalently-closed circles are uncommon in DNA preparations from bacteria, most molecules are either supercoiled, nicked or linear

5   Linear single stranded DNA molecules

- Synthetic oligonucleotides

- DNA molecules produced by copying a template by extension from a fixed primer (5' end) e.g. products of a DNA sequencing or primer extension experiment

- The extent of inter and intra-molecular base pairing is determined by the prevailing conditions

6   Circular single stranded DNA molecules

- Viral genomes: e.g. filamentous bacteriophage M13 (used extensively during *in vitro* mutagenesis for protein engineering)

7   Interlocked double stranded DNA circles

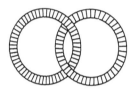

- Interlocked circles are found in the kinetoplast, an organelle unique to protozoa

- Interlocked circles can also be produced in experiments with topoisomerase enzymes which insert supercoils

RNA

8   Single stranded RNA circles

- Produced *in vitro* for experimental tests for the mechanism of the initiation of protein synthesis

9   Double stranded linear RNA molecules

- Viral genomes e.g. *Reovirus*

- Hybrids formed in RNAse protection experiments

10   Single stranded linear RNA molecules

- Viral genomes e.g. *HIV, Rhinovirus*
- tRNA, rRNA, mRNA
- Molecules produced *in vitro* by copying a template from a promoter

- The extent of inter or intra-molecular base pairing (secondary and tertiary structure) will depend on the prevailing conditions

11   Branched single stranded RNA molecules

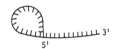

• Intermediates of RNA splicing (removal of introns). One ribose sugar makes both 3'→5' and a 2'→5' sugar–phosphate bonds with two different phosphates. This forms a lariat or branch structure (lariat means a 'lassoo')

RNA: DNA hybrid molecules

12   RNA:DNA hybrids

• RNA:DNA hybrids are formed in nuclease protection experiments such as S1 mapping

of double- and single-stranded molecules, in either circular or linear conformation, are illustrated. RNA molecules and RNA/DNA hybrids are also shown. The second part of the book will consider specific applications of different electrophoretic techniques. Chapter 6 will match the different forms of DNA and RNA molecules shown in *Figure 1.3* to particular electrophoretic applications.

The next chapter reviews theoretical aspects of the migration of nucleic acids in electric fields. Understanding how nucleic acids negotiate their path through a gel often provides the explanation for why a particular method is used for a particular application and, most importantly, the limits and capabilities of the method in question.

References

1. Whitaker, J.R. (1967) *Paper Chromatography and Electrophoresis,* Volume I, Electrophoresis in stabilising media. Academic Press, New York and London, pp. 1–3.
2. Watson J.D., Hopkins, N.H., Roberts, J.W., Steitz, J.A. and Weiner, A.M. (1988) *Molecular Biology of the Gene,* 4th Edn. Benjamin Cummings, Menlo Park.
3. Darnell, J., Lodish, H. and Baltimore, D. (1990) *Molecular Cell Biology,* 2nd Edn. W.H. Freeman, New York.
4. Watson, J.D., Gilman, M., Witkowski, J. and Zoller, M. (1992) *Recombinant DNA,* 2nd Edn. Scientific American Books, New York.
5. Lewin, B. (1994) *Genes V,* 5th Edn. IRL/Oxford University Press, Oxford, p. 1296.
6. Old, R.W. and Primrose, S.B. (1994) *Principles of Gene Manipulation,* 5th Edn. Blackwell Scientific Publications, Oxford.
7. Newton, C.R. and Graham. A. (1994) *PCR.* BIOS Scientific Publishers, Oxford.

2 The Theory of Nucleic Acid Electrophoresis

2.1 The movement of nucleic acids in liquids and in gels

This chapter begins with a thought experiment. If a negatively charged molecule of nucleic acid were placed in a vacuum between a positive and a negative electrode, the nucleic acid would experience a force accelerating it towards the source of the positive charge. The strength of the 'push' would be determined by the voltage and the distance between the two electrodes (see *Figure 2.1*). Voltage divided by distance gives the field strength, which is usually measured in volts per centimeter. Now, consider the results of a real experiment. If

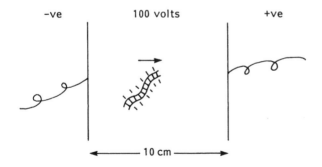

-ve   100 volts   +ve

←———— 10 cm ————→

• Field strength: 10 volts cm^{-1}

FIGURE 2.1: *The force on a charged molecule in an electric field. The voltage provides the driving force on a charged molecule in an electric field. The important parameter is the field strength: the voltage divided by the distance between the two electrodes.*

9

an electric field is applied to the negatively charged nucleic acid in water, there is very little migration towards the positive electrode. Replace the water with a buffer containing positive and negative ions and the molecule is able to migrate. If nucleic acids respond to an electric field in a vacuum, where there is no current flow, why does electrophoresis not occur in water and what happens when charged ions are introduced? The solution to this conundrum lies in the chemistry that takes place at the surface of the electrodes (see *Figure 2.2*).

2.1.1 Electric currents and buffer solutions

In a poorly conductive medium such as water, the few positive and negative ions that are present will move to the electrodes of opposite sign. This effectively screens the nucleic acid such that the field strength at the mid-point between the electrodes is very low. However, in a conductive medium, one with plenty of negative and positive ions, an electrical circuit is completed. (A 0.1 M solution has a concentration of ions 1 million times greater than water at pH 7.) Now electrons can leave the negative electrode. This results in the electrolysis of water to yield bubbles of hydrogen gas and alkali:

$$2e^- + 2H_2O \rightarrow H_2 + 2OH^-.$$

At the positive electrode, electrons can be removed from water to give bubbles of oxygen gas and acid:

$$H_2O - 2e^- \rightarrow \tfrac{1}{2}O_2 + 2H^+.$$

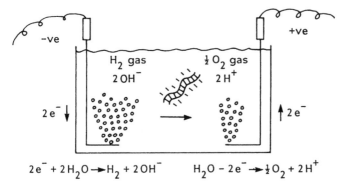

FIGURE 2.2:   *The electrochemical events which take place during nucleic acid electrophoresis. An electric field is established in an aqueous medium when an appreciable concentration of conductive ions are present in the solution. As current flows, water is subjected to electrolysis: the generation of hydrogen gas (H_2) and alkali (OH^-) at the negative electrode (cathode) and oxygen gas (O_2) and acid at the positive electrode (anode). Note the greater number of bubbles at the negative electrode.*

With the occurrence of these reactions there is no screening of the electrodes. The charged nucleic acid experiences the electric field and moves with a force determined by the voltage. Weak acids such as acetate, phosphate, or borate are usually included as the conducting ions during electrophoresis. This has the additional advantage that the OH^- and H^+ ions produced at the electrodes are buffered. This is important, since extremes of pH could denature the structure of the nucleic acid and alter its mobility.

The interrelationship between the flow of current I (the number of charges per second reaching the positive electrode, measured in amps (A)) and the voltage V (the 'push' given to those charges) is described by the equation $V = IR$ (Ohm's law). R is the resistance of a substance to the passage of an electric current. The resistance of a solution decreases as more charged ions are dissolved in it. When charges move through a substance they release energy in proportion to the push (voltage) that they were given. This energy appears as heat. The quantity of heat in joules per second is known as the power (P). Power is measured in watts (W). Power can be calculated by multiplying the current (I) by the voltage (V): $IV = P$. In electrophoretic terminology this generation of heat is often referred to as *Joule heating*.

Although the presence of charged ions is essential for the migration of nucleic acids, too high a concentration of ions spells trouble. In this case the resistance will be low and a high current will pass between the electrodes. The rate of migration of the nucleic acid is dependent on the voltage, so voltage is often maintained at a constant level by the electrophoresis power supply. As $V = IR$ and $IV = P$, substituting IR for V gives $I^2R = P$. Thus, if the voltage is held constant, the consequence of reducing R by increasing ionic strength is a squared increase in the current I. This cannot offset the reduction in power caused by the drop in the resistance and leads to a significant elevation in the amount of heat being generated. This could melt, crack or even set fire to the apparatus! Short of such catastrophic happenings, even modest amounts of heat generated in liquids used for electrophoresis can have undesirable consequences. Heating can set up convection currents, which will tend to mix and circulate the electrophoresis buffer. Since the function of electrophoresis is to separate mixtures of charged molecules, mixing is the last thing that is needed. Thus, the desire to run electrophoretic separations as fast as possible by raising the voltage has to be tempered by the necessity to limit the generation of heat. Reducing the ionic strength will limit heat production but, as we have seen, sufficient ions are needed to complete the electrical circuit. Moreover, it has been discovered that the resolution of separations (how tight the bands are) is improved in higher ionic concentrations. The ionic strengths employed for electrophoresis are therefore a compromise, and are typically of the order of 0.1 M. A readable treatment of the theory underlying the effects of charge, voltage, size, and ionic strength can be found in the early chapters of ref. 1.

2.1.2 Nucleic acids in solution

In the thought experiment at the beginning of this chapter, the velocity of the charged molecule in a vacuum was dependent only on the voltage applied to it. When comparing different molecules, the total number of charges and the mass per molecule would affect the velocity. However, in the real world there are also frictional forces to contend with. Friction between the medium and the charged molecule will act to retard the forward rate of migration. (It is easy to envisage progress being slower in treacle than in a dilute buffer.) The larger the molecule (actually the greater its cross-sectional area) the more severe the retardation. The eventual motion of the molecule then depends on the charge:size ratio. Mixtures of molecules with different charge:size ratios can therefore be separated through their rates of movement in a conducting liquid. The charge on a protein at a particular pH depends on the distribution of charged amino acid side-chains which varies from protein to protein. Proteins can therefore be separated electrophoretically in liquids. Performing electrophoretic separations of proteins in narrow, buffer-filled capillaries is an increasingly important technique, known as *capillary electrophoresis*. The use of a very thin capillary tube, with an internal diameter of less than 0.1 mm, ensures there can be highly efficient cooling to minimize convection currents caused by Joule heating. This permits shorter run times by using high voltages. Capillary electrophoresis can also be carried out in gels as well as in liquids, but it is quite difficult to make gels in very fine capillaries. A discussion of capillary electrophoresis is outside the scope of this book but further information can be found in ref. 2.

In comparison to proteins, nucleic acids are not so obliging. Nucleic acids have a negatively charged phosphate for every base in their sequence. Therefore, they posess a constant charge:size ratio. In consequence, nucleic acids of all sizes migrate at a similar rate in liquids and cannot be separated by this technique. When molecules possess a constant charge:size ratio, methods are required that exploit the frictional or obstructive properties of the electrophoretic medium, so that they can be separated according to their size. For nucleic acids this is the only way in which they may be separated. The most commonly applied solution to this problem is to perform nucleic acid electrophoresis in gels which act as molecular sieves. The same problems occur during the electrophoresis of proteins, when proteins are saturated by a constant quantity of the negatively charged detergent sodium dodecyl sulfate (SDS). The proteins cannot be separated unless they are made to move through a gel. A comprehensive description of the separation of proteins by electrophoretic techniques can be found in the companion volume to this book [3].

2.1.3 Nucleic acids in gels

Gels consist of a solid matrix entrapping a buffer. This results in a series of pores through which the nucleic acid molecules must pass. The effect is like a sieve, or rather like a series of meshes stacked one on top of the other. All the molecules have similar charge:size ratios, so they all move at the same 'speed'. However, the smallest species experience fewer collisions with the matrix, so they pass through more rapidly (see *Figure 2.3*). Two substances are used to make gels for the analysis of nucleic acid molecules: agarose and polyacrylamide. Agarose is essentially highly purified agar. It is a naturally occurring, long-chain polysaccharide obtained from seaweed. Agarose is melted by heating in electrophoresis buffer and forms a gel on cooling. It has relatively large pores, the size of which are determined by the

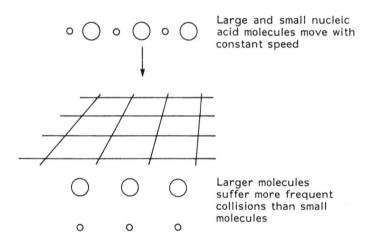

FIGURE 2.3: *The concept of molecular sieving. The constant charge:length ratio of nucleic acids (one negative for each phosphate in the chain) dictates that all species would move at a constant speed in an unrestricted, liquid environment. Separation according to size can only be achieved by passage through a molecular sieve. Larger molecules are retarded since proportionately they suffer more frequent collisions with the sieve matrix.*

concentration of agarose used to form the gel. Polyacrylamide is formed from the synthetic small molecule acrylamide and is polymerized into long chains with the addition of a catalyst. A cross-linker molecule, often methylene-bis-acrylamide (or bis for short), is usually added to bind together the polyacrylamide chains and encapsulate the electrophoresis buffer. Polyacrylamide gels have smaller pores than agarose gels. The size of the pores is determined both by the concentration of the polyacrylamide and the amount of cross-linker. The preparation of agarose and polyacrylamide gels is described in Chapter 4.

To summarize, proteins can be separated in liquids or in gels, but nucleic acid separation is restricted to a gel format. Two consequences of performing electrophoresis in gels as opposed to liquids are well worth remembering. The first is a benefit, the second poses a potential problem. When heat is generated during electrophoresis within a gel, the rigid confines of the gel matrix provide the beneficial effect of inhibiting convection currents. In a liquid, convection currents tend to mix the separating molecules. This is why the electrophoresis of proteins in liquids is invariably performed in an ultra-fine capillary – the high surface area:volume ratio permits very efficient cooling. For gels, a degree of cooling may still be necessary, if sufficient voltage is to be applied to achieve faster run times. Often this is realized by employing a reservoir of electrophoresis buffer, which acts as a heat sink. On the other hand, a potentially detrimental effect of performing electrophoresis in gels is a phenomenon known as *electroendosmosis* (EEO). EEO occurs if there are charges present on the fixed molecules of the gel matrix. Charged water molecules (H_3O^+, a water molecule with H^+ bound) are attracted to fixed, negative charges on the matrix and these water molecules are driven towards the negative electrode. This can provide a counter-current that is sufficient to prevent migration of the negatively charged molecule one is trying to separate. The effect is like trying to paddle a canoe upstream. Agarose and polyacrylamide gel matrices must therefore be as electrically neutral as possible. This is usually achieved by using highly purified electrophoresis-grade components. A value for EEO is often given in the analysis of such chemicals.

2.2 The movement of nucleic acids through gels in constant electric fields

In recent years, there have been a significant number of studies concerned with the production of realistic mathematical models for predicting the behavior of nucleic acids as they move through agarose or polyacrylamide gels [4,5,8–11]. For our purposes, it will be sufficient to describe the results of these studies in the broadest outline. The interested reader can explore this literature by examining the references listed at the end of this chapter.

A combination of theoretical modeling and careful experimentation has shown that nucleic acids can migrate in three different ways in both agarose and polyacrylamide gels. The three formats are:

- Ogston sieving;
- reptation;
- rigid rods.

The form of migration that is adopted depends on the length of the molecule, the size of the pores within the gel and the field strength.

2.2.1 Ogston sieving

Ogston sieving takes its name from a researcher who was amongst the first to describe this particular mode of migration during electrophoresis through a gel matrix [6]. During Ogston sieving, nucleic acid fragments are present in the gel as randomly tumbling, globule-shaped molecules. Passage through the gel is determined by the radius of gyration: the cross-sectional area swept by the molecule as it undergoes random thermal motion (see *Figure 2.4*). The closer the radius of gyration is to the size of the average pore in the gel, the less likely it is that the molecule will pass through a pore in a given time. Very small fragments will migrate almost as if the gel was not present and therefore cannot be separated. At an intermediate size, the more frequent collisions between the gel matrix and molecules with a wider radius of gyration will lead to their separation from slightly smaller species, which will suffer proportionately fewer collisions. Fragments whose radius of gyration is significantly bigger than the cross-sectional area of a pore would never be expected to pass through the gel and should not be separable. However, on closer analysis it appears that nucleic acids can switch from the tumbling globule to migrate in an alternative conformation.

2.2.2 Reptation

A combination of theoretical and experimental studies has established that, under the influence of the electric field, nucleic acids present initially as globules can deform and enter the gel 'end-on'. The beautifully descriptive term *reptation* was coined by De Gennes [7] to describe the reptile-like movement of a linear molecule as it snakes its

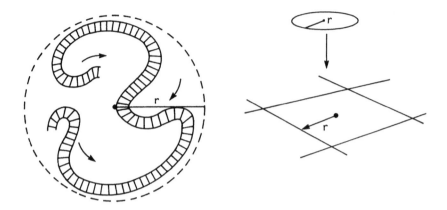

The radius of gyration is the average radius of the area (volume) swept by a molecule as it undergoes random thermal tumbling

Molecules can pass through the gel matrix if they possess a radius of gyration less than that of the matrix pores

FIGURE 2.4: *Migration of nucleic acids by Ogston sieving. Ogston sieving is the term used to describe the process in which nucleic acid molecules are separated according to their size by a gel matrix in which the average pore size is comparable to the radius of gyration for the largest species.*

way through the pores in the gel (see *Figure 2.5*). Of course when it is orientated end-on, the cross-sectional area of the nucleic acid is insignificant, but separation of molecules of different sizes is nevertheless efficient, as the fragments need to wind their way through the matrix. A larger fragment will therefore take longer to move a given distance than a shorter fragment. In polyacrylamide gels single-stranded DNA fragments of several hundred nucleotides in length are efficiently separated by reptation, when they differ by only a single nucleotide. This is of course the basis of DNA sequence determination [8] (see Sections 5.5 and 11.2 for a description of DNA sequencing techniques and applications). It might be argued that this simple but effective gel methodology will make as big an impact on society as any of the discoveries of twentieth century technology. Interestingly, some recent reports have raised the issue of whether the gel matrix becomes realigned under the influence of the electric field. It appears that a measure of matrix reorientation occurs in agarose gels during electrophoresis and this could affect the way in which long molecules reptate through the gel [9].

Refined computer simulations have been developed to investigate the properties of nucleic acids in gels [5]. These theoretical studies have led to a model of migration termed *biased reptation*. The biased reptation model incorporates the biased effect on nucleic acid mobility,

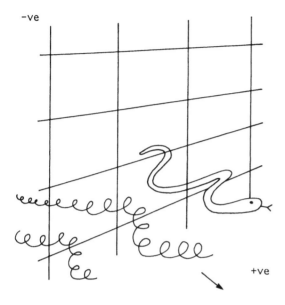

FIGURE 2.5: *Migration of nucleic acids by reptation. Under the influence of an electric field nucleic acid molecules can switch from a randomly tumbling conformation with a relatively large radius of gyration to an elongated molecule that winds its way through the pores of the gel matrix in a process called reptation.*

which derives from the fact that the free ends of a molecule have the ability to switch direction to a much greater degree than an internal segment. Studies based on computer models of reptating DNA molecules suggested that sometimes a longer molecule might adopt a conformation where both ends of the molecule begin simultaneously to snake towards the positive electrode, eventually forming a U-shape (see *Figure 2.6*). Once formed, the U-shaped species may become wrapped around the solid gel matrix, with significantly retarded migration [10,11]. The molecule will only be free to move when one or another of the ends resumes the lead. Computer models predicted this behavior for DNA fragments of an intermediate size. Shorter molecules can readily reorientate themselves out of the U-shape. Longer species very rarely get trapped in this conformation. Subsequent experiments, in which DNA restriction fragments ranging from 4000 to 20 000 bp were separated using agarose gel electrophoresis, showed the exact behavior predicted by the computer simulations (see *Figure 2.7*). At certain concentrations of agarose, molecules of lower molecular weight would run more slowly than significantly larger molecules. This is called *band inversion* and, as its discoverers point out, it is quite probable that as a result of this phenomenon some errors may have occurred in the assignment of molecular weights to genomic DNA fragments detected by Southern blotting [11]. The problem is exacerbated in higher agarose

-ve

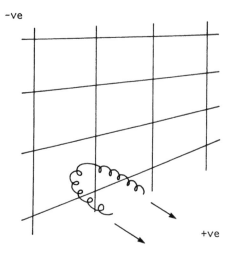

+ve

FIGURE 2.6: *Self-trapping of U-shaped molecules leads to band inversion. Where pore sizes are sufficiently small and field strengths are low, intermediate-sized molecules can become trapped in a U-shaped conformation within the gel matrix. This leads to anomalous properties of electrophoretic migration called band inversion.*

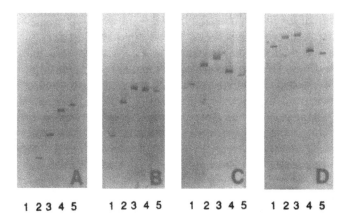

1 2 3 4 5   1 2 3 4 5   1 2 3 4 5   1 2 3 4 5

FIGURE 2.7: *Examples of band inversion. In these experiments double-stranded linear DNA fragments of λ bacteriophage of known sizes were generated by restriction enzyme digestion. These DNA molecules were loaded into separate lanes on gels consisting of different percentages of agarose. Electrophoresis was carried out for 16 hours at 2.4 V cm^{-1}. Lane 1 4.4 Kbp; lane 2 6.6 Kbp; lane 3 9.9 Kbp; lane 4 23.1 Kbp; lane 5 41.7 Kbp. A. 0.8% aagarose, B=1.2% agarose, C=1.6% agarose, D=2.0% agaarose. As the concentration of agarose was increased, pore sizes decreased and the smaller-sized fragments suffered vastly reduced mobilities. Reproduced from ref. 11 with permission from the American Physical Society.*

concentrations and lower field strengths; for example gels subjected to low voltage, overnight runs.

Using constant field agarose gel electrophoresis it is very difficult to separate molecules of sizes greater than 30 000 bp. Indeed, if these very slow moving nucleic acids are subjected to stronger electric fields to encourage their separation, the problem is compounded [12]. Under these conditions, molecules of different lengths display an even greater tendency to migrate with indistinguishable mobilities. The biased reptation computer models have also been employed to explain this effect.

2.2.3 Migration as rigid rods

In reptation, it is proposed that nucleic acids migrate as coiled, helical molecules. However, as field strengths are increased in an attempt to separate larger molecules more quickly, the coiled formation begins to deform as the high voltage stretches the molecule towards the positive electrode (see *Figure 2.8*). Under these conditions the molecule adopts a rigid, rod-like structure. This aids progress through the pores of the gel, but all trace of the retardative properties of the matrix are lost and nucleic acids begin to migrate at size-independent rates. Thus, at sizes considerably above 30 000 bp and at low voltages the molecules might eventually separate, but the gels could take many days to run. If you try to run the gel faster, by turning up the voltage, the separation becomes worse as now the molecules adopt the rigid rod formation. Analyses of the point at which rigid rod migration begins have suggested that nucleic acid deformation will ultimately limit the speed of separation obtainable by using greater voltages in the ultra-fine capillary gels now being researched for the next generation of automatic DNA sequencing machines, even when there is efficient cooling so that the effects of Joule heating can be minimized [8].

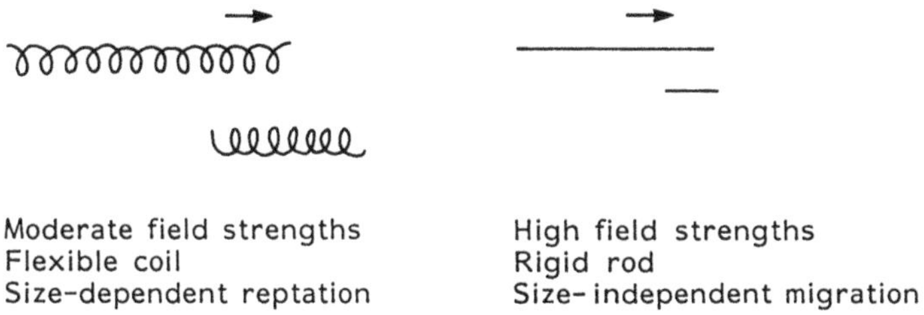

Moderate field strengths
Flexible coil
Size-dependent reptation

High field strengths
Rigid rod
Size-independent migration

FIGURE 2.8: *Migration of nucleic acids at high field strengths. Under moderate field strengths, nucleic acids migrate as flexible coils by size-dependent reptation. At high field strengths, flexible coiled molecules are converted into rigid rods, whereupon migration becomes size independent.*

2.3 The movement of nucleic acids through gels in pulsed electric fields

In the preceding section we saw that nucleic acid fragments above 30 000 bp become quite difficult to resolve in conventional agarose gels. Within the past decade, there has been increasing development of a technique that employs *pulsed electric fields* in place of constant fields. The fundamental principle of pulsed field gels is to use a voltage that turns on and off quite rapidly. Pulsed field gels are able to resolve much larger sizes of nucleic acid. Initially, pulsed field gels involved a simple on–off–on cycle of the electric field in one direction. In subsequent years, the technique has been refined to include switching the pulses to and fro in opposite orientations (a case of two steps forward, one step back), and pulsing the field from different directions. There is insufficient space in this book to delve very deeply into the different varieties of pulsed field gels. Some research applications will be reviewed in Chapter 9. A number of detailed practical guides have recently been published. However, in the light of the present discussion on the conformation of nucleic acids migrating in electric fields, it is appropriate to consider here the effects of temporal and directional alterations in field strength. The biased reptational model [5] has been instrumental in the analysis of pulsed field technology.

Imagine a long DNA molecule migrating within a gel in the reptational mode. If the field is temporarily turned off, the molecule will begin to relax back into its random, globular form (*Figure 2.9*). Depending on when the forward pulse is next applied, the molecule will either have completely refolded, or be in the process of refolding, when it is obliged to readopt the linear reptational conformation. Larger species will spend proportionately more of the time folding and unfolding than smaller molecules. The smaller species will be able to undergo significant forward migration, depending on how quickly they readopt the reptational aspect. Similar arguments apply to pulses coming from different directions. In essence, the pulse frequency can be adjusted or tuned, by trial and error, so that molecules of a certain size range will just be able to fold and unfold in resonance with the pulses. These molecules will make significant forward progress in the gel. In contrast, much larger molecules will never make substantial forward progress, whilst very much shorter molecules will have so much time to fold and unfold that they will migrate rapidly at the leading edge of the gel. Pulsed field gels generally employ very large pore agarose gels. Low voltages are used

Direction of field ⟶

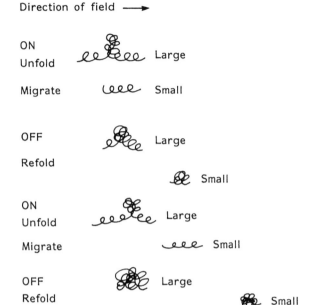

FIGURE 2.9: *Folding and unfolding in pulsed field gel electrophoresis. The understanding that long DNA molecules must adopt an end-on reptational aspect in order to achieve significant forward migration in a gel matrix provides an explanation of how pulsed fields can be used to separate molecules that would fail to separate under constant field gels. The smaller molecule can unfold and begin to migrate before the pulse is turned off. The larger molecule never achieves forward migration since it cannot fold and unfold within the duration of the pulse.*

to prevent the molecules adopting size-independent reptation. Consequently, pulsed field gels involve rather lengthy run times, but separations of double-stranded DNA molecules in the 1 million bp range are now quite routine.

References

1. Whitaker, J.R. (1967) *Paper Chromatography and Electrophoresis,* Volume 1, Electrophoresis in stabilising media. Academic Press, New York.
2. Kuhn, R. and Hoffstetter-Kuhn, S. (1993) *Capillary Electrophoresis: Principles and Practice.* Springer Verlag, New York, p. 300.
3. Dunn, M.J. (1993) *Gel Electrophoresis: Proteins.* BIOS Scientific Publishers, Oxford, p. 176.
4. Slater, G.W., Turmel, C., Lalande, M. and Noolandi, J. (1989) DNA gel electrophoresis: effect of field intensity and agarose concentration on band inversion. *Biopolymers,* **28,** 1793–1799.

5. Noolandi, J., Slater, G.W., Lim, H.A. and Viovy, J.L. (1989) Generalized tube model of biased reptation for gel electrophoresis of DNA. *Science,* **243,** 1456–1458.
6. Ogston, A.G. (1958) The spaces in a uniform random suspension of fibres. *Trans. Faraday Soc.,* **54,** 1754–1757.
7. De Gennes, P.G. (1971) Reptation of a polymer chain in the presence of fixed obstacles. *J. Chem. Phys.,* **55,** 572–579.
8. Grossman, P.D., Menchen, S. and Hershey, D. (1992) Quantitative analysis of DNA-sequencing electrophoresis. *Genet. Anal. Tech. Appl.,* **9,** 9–16.
9. Stellwagen, N.C. and Stellwagen, J. (1989) Orientation of DNA and the agarose gel matrix in pulsed electric fields. *Electrophoresis,* **10,** 332–344.
10. Slater, G.W. and Noolandi, J. (1989) The biased reptation model of DNA gel electrophoresis: mobility vs molecular size and gel concentration. *Biopolymers,* **28,** 1781–1791.
11. Noolandi, J., Rousseau, J., Slater, G.W., Turmel, C. and Lalande, M. (1987) Self-trapping and anomalous dispersion of DNA in electrophoresis. *Phys. Rev. Lett.,* **58,** 2428–2431.
12. Fangman, W.L. (1978) Separation of very large DNA molecules by gel electrophoresis. *Nucleic Acids Res.,* **5,** 653–665.

3 The Electrophoresis of Native and Denatured Nucleic Acids

3.1 The control of base pairing

Intramolecular or intermolecular base pairing can give rise to dramatic effects on the overall shape of a molecule. As an example, consider the intramolecular secondary structure of the 1542 nucleotide sequence of 16S ribosomal RNA from *Escherichia coli* (see *Figure 3.1*). It will come as no surprise that the movement of nucleic

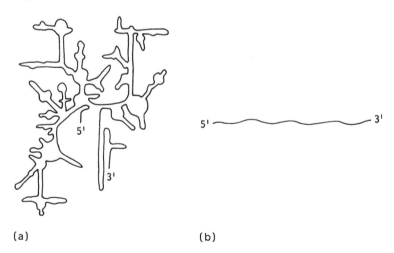

(a)   (b)

FIGURE 3.1: *Nucleic acids under native and denatured conditions. Intramolecular base pairing has a profound effect on secondary structure. This example shows the single-stranded 16S rRNA from* E. coli, *under native (**a**) and denatured (**b**) conditions. An accurate measure of the size of a single-stranded nucleic acid can only be obtained under denaturing conditions. 16S structure after ref. 4.*

acids through the pores of agarose or polyacrylamide gels is profoundly influenced by the degree of base pairing. This may or may not be desirable, depending on the aim of the experiment. In practice, nucleic acid gel electrophoresis is generally performed under either native conditions when double-stranded molecules are to be examined or denaturing conditions for the analysis of single-stranded material. In native conditions natural base pairing is allowed, whilst in denaturing conditions all base pairing is prevented. Sometimes conditions are used where base pairing is permitted to reform in previously single-stranded molecules in order to discriminate between sequences of identical length which differ at one or more nucleotide positions. Under these conditions small nucleotide differences can affect the specific details of intramolecular base pairing and this may permit the separation of two very similar molecules. The most important example of this approach is in the detection of mutations by techniques called single- or double-strand conformational polymorphism (SSCP or DSCP). These topics are covered in Chapter 10, which describes applications of nondenaturing polyacrylamide gels.

Studies have shown that the three forms of nucleic acid migration discussed in Chapter 2 (Ogston sieving, reptation and migration as rigid rods) operate for both single-stranded and double-stranded nucleic acids [1]. The only difference is the point at which molecules switch from Ogston migration as folded globules to 'end-on' reptation. This will occur at shorter molecular lengths for a double-stranded nucleic acid than a single-stranded one, since the double-stranded species will present a slightly larger radius of gyration. Specific research applications involving either native, denaturing or partially denaturing conditions will be reviewed in Part 2. At this point, it is appropriate to survey the various methods that are employed to accomplish nucleic acid denaturation. Further details can be found in refs 2 and 3.

3.2 Physical and chemical denaturants for nucleic acid gel electrophoresis

The different methods by which intramolecular and intermolecular base pairing can be disrupted can be placed into four groups:

- temperature;
- alkaline conditions;
- methyl mercury, glyoxal or formaldehyde;
- urea or formamide.

3.2.1 Temperature

Temperature is the most obvious denaturant for nucleic acids. On forming a base-paired structure, free energy is released. Resupplying this energy, in the form of thermal motion, permits the base-paired regions to disengage. In fact, the stability of base-paired structures is often quoted as T_m, the temperature at which the two strands melt or dissociate. For short stretches of nucleic acid, a rough and ready guide is to allow 2°C for each A–T or A–U base pair, and 4°C for each G–C. (There are two noncovalent hydrogen bonds holding the bases together for each A–T and three bonds for each G–C base pair.) A 20 bp length of DNA containing 10 A–T base pairs and 10 G–C base pairs would therefore melt completely at 10 x 2°C + 10 x 4°C = 60°C. More complicated formulae can be used for accurate calculations or longer sequences (see refs 2 and 3).

In practice, with standard apparatus, it is rather difficult to achieve accurately controlled, elevated temperatures during gel electro-phoresis. Instead, high temperatures, which can be generated through Joule heating, are avoided by ensuring adequate cooling. Agarose gels could start to melt at high temperatures, although polyacrylamide gels are more resilient.

The effect of positive ions. The concentration of positively charged ions also affects the stability of base pairing. This is because when a length of double-stranded nucleic acid is formed, the strength of the hydrogen bonding between the strands has to be sufficient to overcome the natural repulsion between the negatively charged phosphate groups. If the strength from base pairing is insufficient, the molecule splits. In the presence of positive ions, the negative charges can be shielded from each other, and the duplex is stabilized. The concentration of salt [Na⁺] is often quoted for hybridization experiments, when DNA or RNA molecules are detected with labeled nucleic acid probes. Hybrids that persist at low salt or stringent conditions (0.1 x SSC) suggest a high degree of base-pairing homology. Hybrids that are only stable at high salt or relaxed conditions. (2 x SSC) suggests imperfect base pairing and limited homology. (SSC is a standard sodium chloride/sodium citrate buffer used in these experiments).

3.2.2 Alkaline conditions

In preference to thermal denaturation, chemical treatments that disrupt hydrogen bonding are generally used to eliminate nucleic acid base pairing during denaturing gel electrophoresis. The base pairs that hold nucleic acids together in double strands are formed from noncovalent hydrogen bonds. A hydrogen bond is formed when a group which carries either a unit, or a fraction of a unit, negative charge can closely approach a covalently bonded hydrogen atom. The hydrogen represents a localized positive charge. The formation of hydrogen bonds in nucleic acids is shown in *Figure 3.2*.

One means of disrupting hydrogen bonding is to raise the pH. In alkaline conditions, free negatively charged OH⁻ ions interpose themselves between the partners of the nucleic acid hydrogen bond by forming their own hydrogen bonds with the fixed hydrogen atoms.

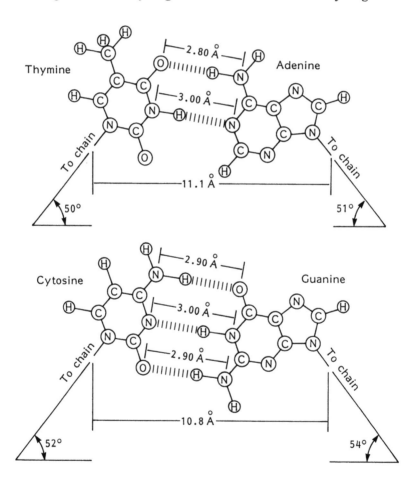

FIGURE 3.2: *Hydrogen bonds formed during base pairing within RNA and DNA. Noncovalent hydrogen bonds are formed between localized sources of negative charge on oxygen and nitrogen atoms in the purine and pyrimidine bases and covalently bound hydrogen atoms.*
Reproduced from ref. 5, Molecular Biology of the Gene, Vol. I, *Fourth Edition, by Watson et al. Copyright (©) 1987 by James D. Watson. Published by The Benjamin/Cummings Publishing Company.*

Nucleic acids are therefore completely denatured into their single-stranded forms. Alkaline agarose gels are used when large (>1000 nucleotides) single-stranded DNA molecules need to be analyzed.

When RNA molecules need to be studied, for example estimating the length of an mRNA species with a Northern blot, it is essential to remove base pairing so that secondary structures do not alter the mobility of the molecule and give a false picture of its size. In this case alkaline denaturation cannot be used. RNA is very much more susceptible to alkaline conditions than DNA. The extra OH at the 2 position of the sugar ribose renders it liable to hydrolysis. When large (>1000 nucleotide) RNA molecules need to be subjected to agarose gel electrophoresis under denaturing conditions, different chemical treatments must be used.

3.2.3 Methyl mercuric hydroxide, glyoxal and formaldehyde: denaturants for RNA in agarose gels

There are three common denaturants compatible with RNA agarose gels: methyl mercuric hydroxide, glyoxal and formaldehyde. All are somewhat volatile and rather toxic. Extreme care must be exercised in their use. These agents interact with the base-pairing nitrogens in purine and pyrimidine rings. Methyl mercuric hydroxide is perhaps the most toxic of the three and many laboratory guides recommend glyoxal or formaldehyde in preference. After gel electrophoresis by each of these methods, brief treatments are used to remove the denaturant from the RNA. This renders the RNA available for base pairing during hybridization, transcription or translation. An application of agarose gels for analyzing denatured RNA is covered in Chapter 8.

Polyacrylamide gels have smaller pores than agarose gels and are capable of greater resolution. When shorter RNAs are to be examined under these conditions, methyl mercuric hydroxide cannot be used, since it reacts with the free radical-generating catalysts used to polymerize the acrylamide monomers. However, both DNA and RNA can be successfully denatured and run on polyacrylamide gels containing formamide or urea.

3.2.4 Urea and formamide

Urea and formamide are small molecules that in high concentrations are capable of disrupting hydrogen bond base pairing. These agents cannot be used in agarose gels because they prevent the gel matrix from solidifying, but they can be used successfully with poly-acrylamide. Either alone, or in combination with formamide, high concentrations of urea completely destroy intramolecular base pairing in RNA or DNA. This is particularly important in sequencing gels, where resolution is required between sequences differing by the length of a single base out of several hundred. Research applications involving high resolution denaturing polyacrylamide gels for single-stranded RNA and DNA are covered in Chapter 11.

3.3 The binding of proteins to nucleic acids during gel electrophoresis

Thus far we have considered the mobility of RNA or DNA during gel electrophoresis in samples devoid of protein. Given our knowledge of nucleic acid migration through the pores of agarose and polyacrylamide gels, it is not difficult to imagine that in nearly all cases the binding of a protein to a nucleic acid will lead to significant retardation in the progress of the molecule. In general, the clarity of electrophoretic separations is improved in deproteinized samples. However, there are occasions where the specific association between a nucleic acid and a protein can give important information on the *in vivo* functions of regulatory molecules with binding sites on RNA or DNA. These experiments are called *gel retardation assays* or *bandshift assays*. Bandshift assays are reviewed in Chapter 10, which describes applications of nondenaturing polyacrylamide gels. At this juncture, it is appropriate to emphasize that in such experiments it is imperative that both the proteins and the nucleic acids are present in their native conformations. This is perhaps self-evident with regard to the protein. A specific binding activity would only be expected in a protein which is folded into a structure that approximates its *in vivo* form. However, by the same token, the binding sites on the DNA or RNA molecules may constitute a specific spatial configuration, such as a stem–loop or a hairpin structure. For DNA, bandshift assays will generally involve a double-stranded molecule. For RNA, the equivalent assay will almost always require single-stranded material. In this instance, it is important to remember that single-stranded RNA may display unexpected mobilities under the conditions for native gel electrophoresis, depending on the degree of intramolecular base pairing that is present. Bandshift assays are thus strictly a qualitative technique. The mobility of a nucleic acid is either affected or not affected, but it may be dangerous to read too much from the degree to which migration is changed.

References

1. Grossman, P.D., Menchen, S. and Hershey, D. (1992) Quantitative analysis of DNA-sequencing electrophoresis. *Genet. Anal., Tech. Appl.*, **9,** 9–16.
2. Sambrook, J., Fritsch, E.F. and Maniatis, T. (1989) *Molecular Cloning: A Laboratory Manual*, 2nd Edn. Cold Spring Harbor Laboratory Press, Cold Spring Harbor, pp. 7.30–7.45.
3. Brown, T.A. (1991) *Molecular Biology LabFax.* BIOS Scientific Publishers, Oxford, pp. 260–261.
4. Moazed, D., Stern, S. and Noller, H.F. (1986) Rapid chemical probing of conformation in 16S ribosomal RNA. *J. Mol. Biol.,* **187,** 399–416.
5. Watson, J.D., Hopkins, N.H., Roberts, J.W., Steitz, J.A. and Weiner, A.M. (1988) *Molecular Biology of the Gene*, 4th Edn. Benjamin Cummings, Menlo Park, p. 149.

4 The Choice of Format: Horizontal or Vertical, Agarose or Polyacrylamide?

4.1 Apparatus for horizontal and vertical gels

Tradition has it that agarose gels are run horizontally, whilst polyacrylamide gels are run in a vertical format. There are no fundamental reasons why this should be so. After all, as we have seen in Chapter 2, the function of the agarose or polyacrylamide matrices is merely to provide different sized holes in the molecular sieve. There are many successful protocols for vertical agarose gels and horizontal polyacrylamide gels. Representatives from electrophoretic equipment companies and electrophoresis enthusiasts lose no time in extolling the virtues of the methods most dear to their hearts. As new techniques and equipment are developed, tradition may be dispensed with. Pharmacia market the PhastSystem™, an automated machine that runs horizontal micro-format polyacrylamide gels. The Multiphore®, also from Pharmacia, is a flat-bed apparatus for a variety of horizontal polyacrylamide applications. Vertical gel electrophoresis with agarose is also beginning to gain popularity. However, in this guide for the newcomer, I will maintain tradition and describe the salient points to consider when purchasing apparatus for running horizontal agarose gels and vertical polyacrylamide gels.

It is common for older guides on gel electrophoresis and molecular biology to reproduce plans and diagrams with which to construct do-it-yourself equipment. Nowadays, there is little to be gained from trying to do experiments with, as a colleague recently put it, 'two old windows and a bulldog clip'. It will generally prove to be a false economy to skimp on the purchase of proper gel equipment. Consider that a busy laboratory might spend in a year, a sum ten times the cost of apparatus, on enzymes and reagents to carry out experiments that ultimately depend on high-quality gel electrophoresis.

It is not my intention to specifically recommend apparatus from particular suppliers. Instead, a few representative products are named in *Figures 4.2* and *4.4* and a check-list of points to watch for in horizontal and vertical gel equipment follows a description of each gel format. A brief guide to electrophoretic power supplies is also included and in the second part of this chapter, the variety of agarose and polyacrylamide media now available is described. Further information on the development of nucleic acid gel electrophoresis can be found in Chapter 6 of ref. 1.

4.1.1 Apparatus for horizontal agarose gels

The essential features of an apparatus for running horizontal agarose gels are illustrated in *Figure 4.1*. All manufacturers use transparent perspex (known also as Plexiglas™) in the construction. This allows the process of electrophoresis to be monitored visually, without disassembling the apparatus. Agarose gels are cast as rectangular slabs by pouring molten agarose solution into an open mold and allowing it to set. Gels are usually around 5–10 mm thick. Most laboratories use a domestic microwave oven to melt the agarose powder and dissolve it into the gel buffer. The wells into which the samples will be loaded are formed by placing 'combs' into the gel before it solidifies. The most convenient method is to use an apparatus in which the gel is formed in an open tray. The tray should be made of a material that is transparent to ultraviolet (UV) light. This permits the migration of nucleic acids to be monitored during electrophoresis if a nucleic acid-binding dye, which fluoresces under UV light (usually ethidium bromide), is incorporated into the gel or the buffer (see Sections 5.3 and 7.5). The ends of the tray should be sealable by some removable device or casting apparatus. This forms the mold into which the molten agarose is introduced. Reject any equipment that requires meters of adhesive tape to effect an efficient seal. You should acquire a variety of combs to allow different numbers of samples to be analyzed. Somehow, wider bands look more pleasing, although it is often useful to be able to run a larger number of samples, whereupon you will be happy to trade the photogenic qualities of your gel for an increase in capacity. A useful way of doubling the number of samples that can be run on a single gel is to make two (or more) sets of wells,

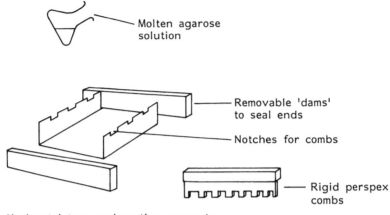

Horizontal tray and casting apparatus

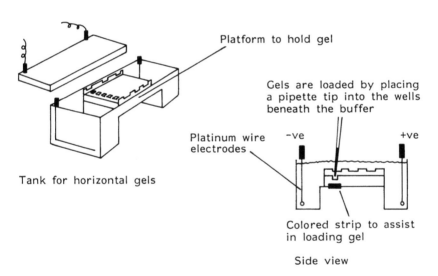

Tank for horizontal gels

Side view

FIGURE 4.1: *Essential features of apparatus for horizontal agarose gel electrophoresis. There are many variations on this general theme. It generally proves useful to invest in a spare casting assembly. In this way one gel can be setting whilst another is being run. It is essential to pour and run gels as horizontally as possible. Failure to observe this can lead to inconsistent nucleic acid separation across the width or the length of the gel.*

by inserting a second comb halfway down the gel. Of course you must be careful not to run the samples too far! Loading the samples into the wells is assisted by a colored strip beneath the apparatus or the gel tray. This makes it easier to see the empty well to be filled.

The different techniques that use horizontal agarose gels will often be appropriate to equipment of different dimensions. Routine electrophoresis to check progress in plasmid construction or PCR reactions can be performed on small gels, saving time and materials. More exacting applications, such as separating complex mixtures of eukaryotic DNA or RNA for Southern or Northern blotting, often work better in the larger format. *Figure 4.2* illustrates the approximate sizes and names of small- and large-format gels from several suppliers. (A list of addresses for suppliers of equipment and consumables appears in the Appendix.)

The electrophoresis tank has a raised platform on to which the gel slab is positioned. It is most convenient if the gel can be retained in the casting tray with the ends free of the device used for sealing the mold. A single electrophoresis buffer is used to fill the tank and cover the gel to a depth of a few millimeters. This prevents the gel from drying out and helps to cool the gel during electrophoresis. The gel is prepared by melting the powdered agarose in the same buffer used to fill the tank (it is best if the same batch of buffer is used since small differences in the concentrations of ions can adversely affect performance). The conductivity of the gel and its shallow covering of

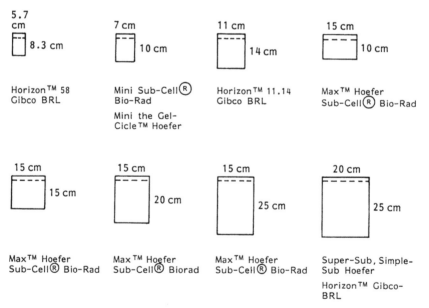

FIGURE 4.2: *A selection of horizontal agarose gels of different sizes. The sizes and dimensions of a few representative products are illustrated to convey the different dimensions that are used for agarose gel. Smaller gels use less agarose, less buffer and give faster separations. Larger gels can be important for resolving large numbers of closely spaced molecules. Gels are drawn to scale.*

buffer are therefore almost identical. If the voltage is maintained at a constant value by the power supply, migration of the nucleic acid will be unaltered by small differences in the depth of the buffer over the gel. However, if too much buffer is added the field strength will drop and electrophoretic migration will slow down. Moreover, extra current will be drawn and this will lead to increased heat generation, although to some extent this is countered by the increased cooling power of the additional buffer.

Passing an electric current through a tank of a conductive solution does of course represent a hazard to anyone foolhardy enough to dip their fingers in whilst the power is on! To eliminate this possibility, insist on an apparatus where there is a lid, to which the positive and negative power cables are attached. It is thus impossible to get at the tank and touch the gel, without disconnecting the power. If the lid is transparent, progress can be monitored through the use of colored marker dyes which migrate in front of all but the smallest nucleic acids (see Section 7.4).

In Section 2.1 we reviewed the electrochemical events that take place during electrophoresis. At the positive electrode, acid and bubbles of oxygen gas are produced. At the negative electrode, there is evolution of hydrogen gas and alkali. For many electrophoresis buffers, the production of acid and alkali is not a problem during runs of moderate duration. The buffering capacity of the solution is sufficient to absorb H^+ and OH^- production, without allowing pH to vary. However, for some buffer solutions and for extended runs the buffering capacity may be exhausted, and pH could rise and fall dramatically at opposite ends of the tank. To compensate for this, some protocols advocate circulating the buffer with a peristaltic pump, from one end of the tank to the other. To allow for this possibility, most manufacturers of electrophoresis apparatus, particularly the larger gel formats, incorporate ports into the lid for tubing with which to effect the recirculation. In general, if the gel tank you are using contains an appreciable volume of buffer, the voltage is not too high and the run is not too lengthy, pH changes should not present a problem. If you suspect pH changes during an electrophoresis experiment, it would be wise to monitor pH intermittently, with a pH electrode, during a control run in the absence of valuable samples. Turn the power off before making the pH reading! In practice, mixing the contents of the tank by lifting out the gel, pouring all of the buffer into a large beaker and returning it to the tank is an effective substitute for recirculation via a peristaltic pump, but not one which you would be happy to repeat at regular intervals during an overnight run! A check-list of the points to consider in the acquisition of apparatus for horizontal agarose gels follows.

- Ease of assembly and trouble-free casting without resort to adhesive tape.
- An ability to cast gels outside the gel tank. This permit gels to be cast in advance, whilst the unit is in use.
- Effective cooling through a heat sink or an adequate buffer reservoir.
- Buffer volume sufficient to limit pH variations.
- Ports for buffer recirculation.
- Precautions to prevent risk of electric shock.
- Ease of sample loading: colored strips beneath the wells.
- Ability to transport gels on tray of UV-transparent material.
- Ability to use commercially produced 'ready-made' gels.

4.1.2 Apparatus for vertical polyacrylamide gels

Those readers who already have experience with the electrophoresis of proteins will feel quite at home with nucleic acid electrophoresis on vertical polyacrylamide gels. The principal components of the apparatus are identical (see *Figure 4.3*). The acrylamide mixture is 'catalyzed', usually by the addition of ammonium persulfate and an amine-containing compound: N,N,N',N'-tetramethylethylenediamine (TEMED). The operator then has anything from 5 to 20 minutes to pour the mixture into the mold before it sets. The space in the mold is formed by clamping plastic 'spacers' between two plates, at least one of which is made of glass to allow the progress of electrophoresis to be observed. The base of the gel mold is formed either by the addition of a third spacer, a length of adhesive tape, a plug of polyacrylamide introduced prior to pouring the gel, or for small gels by simply resting the mold on a flat surface. The cross-section of the gel is determined by the thickness of the spacers. The wells into which the samples are introduced are formed in the top of the gel by the introduction of traditional 'square tooth' combs, or alternatively by using 'sharks-tooth' combs (see *Figure 4.3*). Sharks-tooth combs have teeth on one side and a flat edge on the other. Initially, these combs are inserted upside down whilst the gel is setting. They are removed and then reinserted, teeth down, once the gel is set. Sharks-tooth combs are especially useful for high-resolution nucleic acid sequencing gels, as they create tightly adjacent lanes. This greatly facilitates the comparison between tracks where the bands contain molecules that differ by a single base. The application of high-resolution denaturing polyacrylamide gels for techniques such as nucleic acid sequencing, footprinting or primer extension will be discussed in Chapter 11.

Despite the overall similarity between polyacrylamide gels for nucleic acids and proteins, there is one important aspect that divides them.

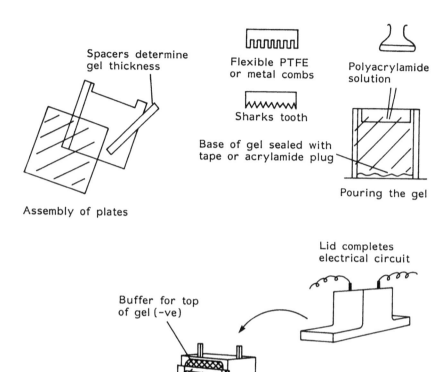

Spacers determine gel thickness

Flexible PTFE or metal combs

Sharks tooth

Polyacrylamide solution

Base of gel sealed with tape or acrylamide plug

Pouring the gel

Assembly of plates

Lid completes electrical circuit

Buffer for top of gel (−ve)

Buffer for base of gel (+ve)

Running the gel

Clamps hold gel sandwich together

FIGURE 4.3: *Essential features of apparatus for vertical polyacrylamide gel electrophoresis. There are many variations to the design illustrated here. Most differences derive from alternative solutions to the problem of separating the buffer chambers for the top (−ve) and the base (+ve) of the gel without allowing the current to short-circuit the polyacrylamide sandwich. There are also a variety of options when it comes to providing cooling for excessive Joule heating in the larger gel high-voltage formats. Uneven gel temperatures can lead to distorted separations: migration is faster in high temperatures so if gels are hot in the center and cool at the edges a 'smile' pattern develops. This can complicate the comparison between nucleic acids in different lanes.*

In the most widespread protocol for the denaturing gel electrophoresis of proteins (SDS–PAGE), gels are poured in two sections. The lower gel, poured first, functions as the molecular sieve or separating gel and contains Tris buffer at pH 8.0. On top of this is a large-pore 'stacking gel', containing Tris buffer at pH 6.8. It is into this stacking gel that the well-forming comb is placed. The gel running buffer, which forms the electrical contact between the positive and negative electrodes, contains Tris and glycine. This arrangement of gels and buffers is known as a discontinuous system and it offers improved resolution for protein gel electrophoresis. (A detailed description of how a discontinuous buffer system functions can be found in ref. 2.) In contrast, discontinuous buffer systems are not widely used for the electrophoresis of nucleic acids, although specialists in methods development are experimenting in this direction. Instead, a single buffer is used to form the gel and make contact with the electrodes in most protocols.

As with horizontal agarose gels, it is often appropriate to run vertical polyacrylamide gels of different sizes for alternative applications. For some techniques involving the nondenaturing analysis of short double-stranded molecules, such as restriction digests, PCR products or nuclease protection assays, a small-scale apparatus identical to that used for proteins is ideal. However, in high-resolution denaturing gel electrophoresis, for applications such as nucleic acid sequencing or footprinting, much longer gels than those common in protein electro-phoresis are needed. These techniques often require gels of 40–50 cm in length, to enable molecular sieving to resolve strands differing by just a single base. Gels that need to achieve this resolution are run at 1500–2000 V. Needless to say, for the careless worker this represents quite a considerable risk of electric shock, especially with home-made apparatus. When purchasing commercially produced equipment, ensure that the apparatus cannot operate with the buffer chambers exposed. Nevertheless, it is common practice to use 'staggered' loading with this sort of gel, since it is possible to load up to 96 samples on some formats. In staggered loading, a few lanes are loaded and the power is turned on for just a minute or so, to run these samples into the first few millimeters of the gel. The power is then turned off, the next set of samples are loaded and the power is turned on again. This procedure minimizes sample diffusion out of the wells and enables a large number of samples to be loaded. Of course, samples whose mobility has to be compared exactly need to be loaded together. Therefore, it pays to look for a system where access to the top of the gel is simple and rapidly achieved. *Figure 4.4* illustrates the sizes of small-, medium- and large-format vertical gels from a range of suppliers. A check-list of the points to consider in the acquisition of apparatus for vertical polyacrylamide gels follows.

- Ease of assembly and casting of gels without resort to excessive lengths of adhesive tape.

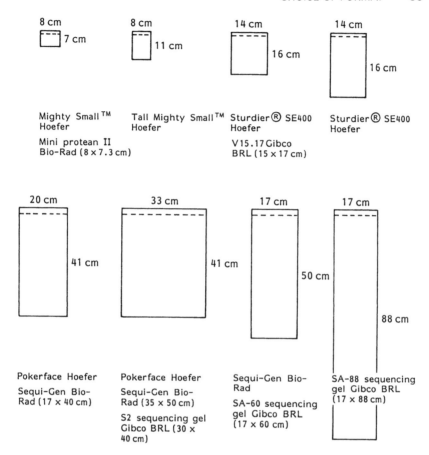

FIGURE 4.4: *A selection of vertical polyacrylamide gels of different sizes. The sizes and dimensions of a few representative products are illustrated to convey the different dimensions that are used for polyacrylamide gel electrophoresis. Smaller gels use less polyacrylamide, less buffer and give faster separations. Larger gels are essential for resolving large numbers of closely spaced molecules. Gels are drawn to scale.*

- Ability to cast gels outside the gel tank permitting gels to be prepared whilst the unit is in use.
- Effective cooling either by buffer reservoir, heat conductive materials or cooling circuits for tap or thermostatically controlled water circulation.
- Precautions to prevent risk of electric shock.
- Ease of sample loading: rapid access for staggered loading in DNA sequencing formats; good visualization of sample wells.
- Siphon or drain to assist in the disposal of any radioactively contaminated buffer.
- Ability to use commercially produced 'ready-made' gels.

4.2 Power supplies for nucleic acid gel electrophoresis

There is a large variety of power supplies suitable for gel electrophoresis on the market. I offer some points to consider when selecting equipment.

- Do not be tempted by the desire to find one power supply which can do everything. It is inflexible to have to run all your gels in one corner of the laboratory.
- Digital LED displays are easier to set than knobs and dials.
- Power supplies should be able to deliver constant voltage or constant current. Constant power and volt–hours integration is also useful.
- Insist on a power supply with a timer so that a gel can be switched off automatically whilst you are out of the laboratory.
- Check that the maximum voltage and current that can be delivered is sufficient for all your applications. High-resolution polyacrylamide gels used for DNA sequencing or footprinting may require 2000 V. If you plan to electroblot protein gels (Western blotting) you may not be able to use the same power supply. Electroblotting requires a large current (a minimum of 500 mA, often more), but a low voltage. Few power supplies will be able to satisfy both these requirements.
- Check that very low settings can be accommodated. Some LED power supplies may only work in increments of 20 V or more. This may be too high for some applications such as electroblotting or overnight gel runs.

4.3 Agarose gel media for nucleic acid electrophoresis

Agarose is a long-chain polysaccharide containing repeating units of D-galactose and 3,6-anhydro-L-galactose (see *Figure 4.5*). The material is extracted from seaweed and inevitably contains a variety of polysaccharide, sulfate and protein contaminants. The level of charged matrix compounds giving rise to EEO (see Section 2.1) should likewise be minimal. Since the method used most frequently to detect DNA and RNA within agarose gels is staining with ethidium bromide, a low level of background fluorescence is also desirable. Clearly a

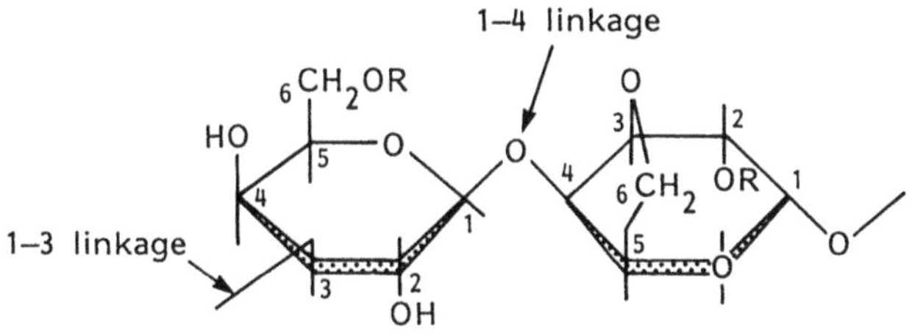

1,3-linked β-D-galactopyranose   1,4-linked 3,6-anhydro-α-L-galactopyranose

FIGURE 4.5: *The molecular composition of agarose. Agarose is a descriptive name for a family of different polysaccharides with the basic repeat structure as shown. Different properties result from the presence of substituents at R. These are obtained either by extraction from different species of seaweed or by chemical treatment. Note that agarose is not a branching or cross-linked molecule.*

product devoid of RNase and DNase activity is required. Products specified as suitable for molecular biology are likely to satisfy each of the above criteria.

Electrophoresis through agarose gels is not used solely to provide analytical information. Once the band corresponding to a particular molecule has been identified, recovery of DNA and RNA from the gel is the most common preparative technique in molecular biology. There are a large number of methods available with which to recover nucleic acid from agarose gels. These include: electroelution into dialysis bags; absorption onto positively charged DEAE (*Di*ethyl*a*mino*e*thyl) paper; absorption onto resin or glass matrices in fine suspensions, columns or on filters; digestion with agarase enzymes. Whichever method is employed, an important factor in the successful use of gel-purified nucleic acids in molecular biology is the level of sulfated polysaccharide contaminants that co-purify with the DNA or RNA and which act as potent inhibitors for polymerase, ligase or restriction enzymes (these inhibitors are present in abundance in agar used for bacterial plates). Molecular biology folklore has it that the presence of inhibitors varies significantly, even between different batches of the same product. Many companies now offer (at a premium) grades of agarose that are tested for their compatibility with gel purification of nucleic acid molecules and the successful manipulation of the purified material in subsequent procedures. An interesting guide to the use and production of agarose products has been produced by FMC BioProducts [3].

4.4 Modified agarose preparations

The demand for highly purified agarose in molecular biology has led to the introduction of agarose preparations with a number of useful alterations to the standard range of properties.

- Low melting point agarose.
- Higher strength agarose.
- High-sieving agarose.
- Visigel™ separation matrix.

4.4.1 Low melting point agarose

An alternative method for extracting DNA from agarose gels prior to further treatment with polymerases, ligase or restriction enzymes is simply to remelt the gel and perform these manipulations in the molten agarose solution. Whilst standard agarose sets at approximately 35°C, it requires heating to between 80 and 90°C in order to melt it. This will be sufficient to dissociate the double-stranded nature of duplex DNA. Moreover, in order to keep the gel molten, enzyme manipulations would have to be performed at temperatures in excess of 40°C. Many manufacturers have therefore developed low melting point agarose specifically for this application. Low melting point agarose is formed by the chemical modification of standard agarose to introduce hydroxyethyl groups into the polysaccharide chain. These modified agarose products gel at temperatures below 30°C but, most importantly, they can be melted at 65°C, a temperature that will not dissociate duplex DNA molecules unless they are extremely short. Molecular biology can therefore be performed *in situ* by melting a slice of agarose containing your band of interest, cooling to 37°C and adding the appropriate enzymes to the still liquid solution. It is of course vital that polysaccharide enzyme inhibitors be absolutely minimal in low melting point agarose used for these applications. It is however, prudent to reduce the potential for enzyme inhibitors by using the lowest concentrations of agarose compatible with the successful resolution of the required fragments. However, it is important to remember that low melting point agarose gives poorer resolution than standard agarose and has a significantly lower mechanical strength.

4.4.2 Higher strength agarose

The average size of the pore in an agarose gel matrix is determined by the concentration of agarose in the gel. Applications involving the separation of very large molecules therefore require large pores and relatively low-percentage gels. Typically, for medium-sized molecules, agarose gels are used at around 0.8–1%. As the concentration of agarose falls to 0.4% agarose gels become very fragile and are easily broken. One solution is to pour low-percentage agarose gels on top of a higher concentration agarose gel base containing the same buffer. The double-layer gel is now more robust and manageable in subsequent manipulations. An alternative approach adopted by some manufacturers is to modify the extraction procedure to refine a fraction of agarose with a higher mechanical strength at low concentrations. These products are intended particularly for pulsed field gel electrophoresis applications (described in Chapters 2 and 9), where very large pore agarose gels are required. Mechanical strength values are usually quoted as grams per square centimeter, the weight that can be supported by 1 cm^2 of a 1% gel. Standard gels score around 1000–2000 g cm^{-2}. Low melting gels have values around 200 g cm^{-2}, whilst high-strength gels can support up to 6000 g cm^{-2}.

4.4.3 High-sieving agarose

In comparison to polyacrylamide, agarose gels are much easier to prepare. Polyacrylamide gels are formed by the addition of a catalyst to a specific mixture of the acrylamide monomer and a bifunctional cross-linker. Oxygen, which inhibits the polymerization, needs to be excluded and, to cap it all, the monomer and cross-linker are rather toxic. The drawback with agarose is that the relatively large pore sizes mean that there is very limited resolution of molecules below 1000 bp. It is not surprising therefore that efforts are being made to produce agarose with greater sieving properties to extend the range of molecules that can be resolved without resorting to the inconvenient nondenaturing polyacrylamide gel. FMC Bioproducts have recently developed a high-sieving agarose named Metaphor. A 4.5% Metaphor gel is capable of resolving the common three-base deletion mutation causing cystic fibrosis (*ΔF508*) from the wild-type allele on PCR fragments of 94 and 97 bp. The agarose is handled in a conventional fashion, except that once it has set it must be cooled to 4°C to develop its high-sieving properties. (Note that some protocols for high-sieving agarose advocate using these gels in a vertical format. Agarose gels can generally be poured and run without too much difficulty in apparatus used for conventional protein gels.) It should be pointed out however, that high-concentration agarose gels take a long time to run (see Section 7.2).

4.4.4 Visigel™

The company Strategene have recently launched a high-sieving high-clarity media under the name Visigel™. An agarose-based preparation containing proprietary additives, this matrix has been formulated to resolve nucleic acids of 700–3000 nucleotides, for example PCR fragments. The material is supplied as an agarose slurry in isopropanol, and gels are prepared by heating with buffer in a microwave as usual. The resulting gel has a much greater transparency than conventional agarose, which when used at concentrations sufficient to resolve nucleic acids in the 700–3000 bp range (2–4%) is rather opaque and low concentrations of ethidium-stained material are difficult to detect. Promotional literature also claims Visigel™ is strong enough to drop without breaking. Whilst I have not put this to the test, conventional agarose gels would certainly not survive such treatment.

4.4.5 Ready-made agarose gels

A number of companies have introduced plastic wrapped, ready-made agarose gels: the molecular biology equivalent of the TV dinner. Whilst time is money, when agarose gels are simplicity itself to prepare, it is difficult to see ready-made agarose gels replacing home-made ones on a wide scale.

4.5 Polyacrylamide gel media for nucleic acid gel electrophoresis

---

Polyacrylamide gels are formed by polymerizing the short monomeric acrylamide molecule in the presence of a variable quantity of a double-ended bis-acrylamide cross-linker (see *Figure 4.6*). Polymerization is catalyzed with the generation of free radicals provided by ammonium persulfate and stabilized by the compound TEMED. This leads to the formation of chains of acrylamide monomers, with occasional cross-links provided by the bis-acrylamide connector. Polyacrylamide pore sizes can be altered in two ways. As the overall concentration of total acrylamide (monomer plus cross-linker) increases, pore size decreases. The value for total acrylamide concentration is given as %T. Experiments have shown that the smallest pore sizes are obtained when the cross-linker is 5% of the weight of the total acrylamide. This is written 5%C. At a value of %C above and below 5%, pore sizes increase. The polymerization of polyacrylamide gels is an exothermic process: it generates heat. If polymerization is too rapid, that is if too much ammonium persulfate and TEMED are added, heating will give rise to convection currents as the gel solidifies and lead to the formation of a nonuniform gel. Oxygen inhibits the polymerization process, so gels must be poured between plates to exclude oxygen from their surfaces (see *Figure 4.3*). Gel mixtures are sometimes 'degassed' by holding under a vacuum for a short period before polymerization.

4.5.1 Modified poylacrylamide formulations

Figure 4.6 illustrates the formation of a standard acrylamide:bis-acrylamide gel, where the compound N,N'-methylene-bis-acrylamide acts as the cross-linker. Other cross-linkers are available and this seems to be an area where future developments may arise. *Figure 4.7* shows some alternative cross-linkers and their effects on the properties of polyacrylamide gels. Of particular interest is the formation of dissolvable gels using the cross-linker N,N'-bis-acrylylcystamine (BAC). This cross-linker can be cleaved by the addition of reducing agents to the gel slice. Since polyacrylamide is a completely synthetic matrix, it is not contaminated with the kinds of natural product enzyme inhibitors that plague agarose gels. DNA extracted from standard polyacrylamide gels is very clean and biologically active, but recovery is often troublesome. The ability to dissolve the gel when constructed with BAC cross-linker makes this technique particularly attractive, although one must ensure that nucleic acids are not damaged in the process.

The company AT Biochem have developed a number of new polyacrylamide formulations marketed under the name Hydrolink Gels. (Please note AT Biochem was sold to FMC BioProducts in 1995.) The matrices have been conceived with the aim of achieving improved electrophoretic performances over standard gels. Long Ranger™ gels offer greater strength and resolution in DNA sequencing applications. Mutation detection enhancement (MDE™) gels are used in DSCP or SSCP detection (see Chapter 10). PCR Purity Plus™ gels offer increased resolution of PCR products. Whilst the developers of these products are keeping rather tight-lipped about the exact composition of their products, gels of the Hydrolink family contain a chemically modified acrylamide monomer, a co-monomer and novel proprietary cross-linkers. Hydrolink gels are more accurately classified as vinyl polymers than polyacrylamide gels.

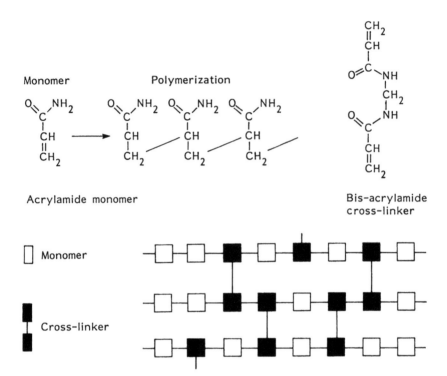

Polymerization of acrylamide and bis–acrylamide cross–linker

FIGURE 4.6: *The polymerization of acrylamide monomers into polyacrylamide gels. On initiation with TEMED and ammonium persulfate, acrylamide monomers polymerize into long chains. Occasionally the bifunctional bis-acrylamide cross-linker is incorporated into the chain and facilitates the formation of a large branching polymer.*

4.5.2 Ready-made polyacrylamide gels

Ready-made polyacrylamide gels are supplied by several companies. Indeed, they are essential for automated gel systems such as the PhastSystem™ from Pharmacia. Since polyacrylamide gels are a little more onerous to prepare than agarose gels, the convenience of an off-the-shelf gel could be attractive if running such gels is likely to be a relatively infrequent event. This would be particularly true for gels that contain a gradient in %T. In gradient gels the pore size gets progressively smaller as the sample moves down the gel. This extends the range of molecular sizes that can be examined on a single gel. Gradient gels find extensive use in the gel electrophoresis of proteins, but they are not commonly used for nucleic acids. It can be rather tricky in the laboratory to produce consistent gradients, be they concave, convex or linear. Commercially produced gradients represent a more uniform source of gels.

Reduced background for
silver staining

Increased resolution

Increased gel strength

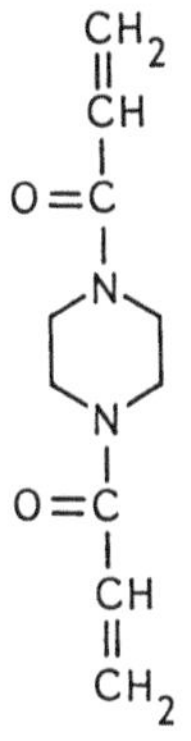

CH$_2$
‖
CH
|
O=C
|
(piperazine ring structure with two N atoms)
|
O=C
|
CH
‖
CH$_2$

Piperazine diacrylamide (PDA)

Gels can be solubilized with
2-mercaptoethanol or
dithiothreitol

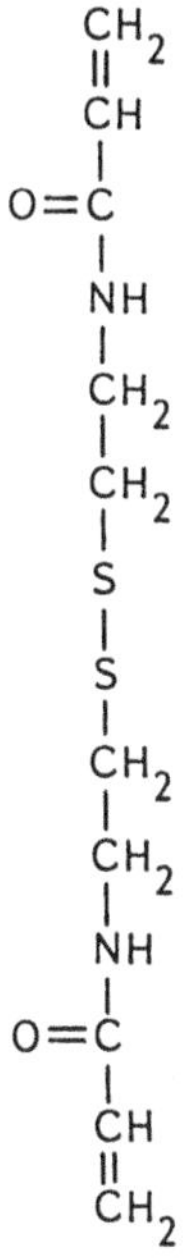

CH$_2$
‖
CH
|
O=C
|
NH
|
CH$_2$
|
CH$_2$
|
S
|
S
|
CH$_2$
|
CH$_2$
|
NH
|
O=C
|
CH
‖
CH$_2$

N,N'-bis-acrylylcystamine (BAC)

FIGURE 4.7: *Alternative cross-linkers for polyacrylamide gels. The use of alternative cross-linkers to the traditional bis-acrylamide (N,N'-methylene-bis-acrylamide) can give different properties to polyacrylamide gels.*

4.5.3 Wedge shaped gels

An alternative to gradients of matrix components, which allow a greater range of molecular sizes to be resolved on a single gel, is to set up a gradient in field strength. This is accomplished by using a wedge shaped gel. The gel is thin at the point of loading but gradually thickens towards the base. The field strength drops as the cross section of the gel widens, and the migration of molecules slows down. This has the effect of preventing shorter fragments running off the end of the gel. Wedge shaped gels are sometimes used for high resolution denaturing polyacrylamide gels to accommodate a larger number of bands in DNA sequencing applications. The gel is formed by using wedge shaped spacers which taper from 2.00 mm to 0.40 mm in thickness.

4.5.4 Toxicity of the components of polyacrylamide gels

Finally in this chapter, it is worth emphasizing once again that the acrylamide monomer and the bis-acrylamide cross-linker are toxic. The danger with these substances is that they have cumulative effects on the nervous system. The polymerized gel is harmless but may contain trapped, unpolymerized monomers. The wearing of disposable gloves is therefore essential when working with acrylamide solutions and polyacrylamide gels. The greatest danger is presented by weighing out the extremely light and airy powders. A welcome advance is therefore the provision, by most suppliers, of ready weighed and mixed acrylamide:bis-acrylamide solutions or powders. This eliminates the most hazardous aspect of their use and is recommended.

References

1.   Sambrook, J., Fritsch, E.F. and Maniatis, T. (1989) *Molecular Cloning: A Laboratory Manual*, 2nd Edn. Cold Spring Harbor Laboratory Press, Cold Spring Harbor, pp. 6.3–6.60.
2.   Dunn, M.J. (1993) *Gel Electrophoresis: Proteins*. BIOS Scientific Publishers, Oxford.
3.   FMC BioProducts Catalog 1995. Technical applications, pp. 68–110.

5 The Detection of Nucleic Acids Following Electrophoretic Separation

5.1 Overview

No information can be obtained from an electrophoretic separation unless the positions of the nucleic acids can be detected and recorded. Whilst the focus of this book is the theory and practice of nucleic acid electrophoresis, it would be a serious omission not to review the variety of methods that exist for detecting DNA and RNA molecules once an electrophoretic separation has been achieved. This is an area in which rapid changes have taken place in recent years. In particular, there has been a marked shift away from radioactive detection to nonradioactive methods. There is insufficient space to cover this material in any great detail. A full description of methods for nucleic acid detection might properly fill another book in this series. This chapter can provide only a broad overview of the different strategies one can adopt.

In Chapter 1, I advanced the notion that, by and large, the various types of nucleic acid gel electrophoresis were not competing alternatives, favored by one laboratory or another. Instead, different electrophoretic methods are geared towards the analysis of different types of molecules, and with different resolutions. To some extent the reverse is true of detection methods. Whilst certain electrophoretic analyses demand particular strategies for nucleic acid detection, there are often several alternatives to choose from. In many cases, individual preference and previous successful experience are the factors that dictate which methods are employed by particular laboratories.

5.2 The principles of nucleic acid detection

There are five principal techniques for detecting nucleic acids:

- binding of fluorescent dyes;
- labeling with radioactive nucleotides;
- labeling with fluorescent nucleotides;
- labeling nucleic acids with specific proteins;
- conventional staining.

Each of these methods will be explained in more detail below. There are, in addition, two principal approaches for nucleic acid detection:

- detection *in situ*;
- detection after transfer to membranes.

It is important to remember that nucleic acids can be detected either by intrinsic methods, the direct observation of the nucleic acid in which you are interested, or through hybridization with a 'probe'. A probe is a labeled, complementary DNA or RNA molecule whose sequence permits it to pair with homologous nucleic acids and thus reveal their position. Following an examination of the five principal methods of nucleic acid detection, I describe which experimental approaches call for detection *in situ* or detection following transfer to membranes, and indicate which methods use intrinsic or hybridization-based techniques.

5.3 Binding with fluorescent dyes

There are a number of dyes that possess the useful property of low fluorescence when free in solution, but high fluorescence when bound to DNA or RNA. The most widely used compound is ethidium bromide. In common with many nucleic acid dyes, this molecule binds to nucleic acids by intercalating between adjacent bases in the DNA or RNA strand (see *Figure 5.1*). Intercalation, the interposition of the flat planar rings of the dye between the flat planar rings of the purine and pyrimidine rings, is more efficient for double-stranded than single-stranded nucleic acid. In this position the molecule is able to absorb UV wavelengths of light and re-emit this energy in the visible spectrum (560 nm). DNA and RNA are illuminated as red–orange bands on a dark, largely nonfluorescent background. UV radiation at 254 nm is absorbed strongly by nucleic acid. This energy is absorbed by the purine and pyrimidine bases, and transmitted to the bound

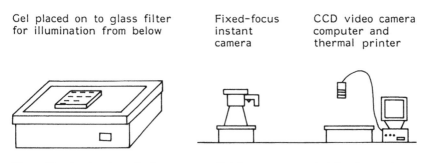

Ethidium bromide

Intercalation of ethidium bromide into double-stranded DNA

YOYO-1

TOTO-1

YOYO-1 and TOTO-1 bind irreversibly to nucleic acids

Gel placed on to glass filter for illumination from below

Fixed-focus instant camera

CCD video camera computer and thermal printer

Transilluminator irradiation with 302 nm ultraviolet

Gel images recorded by instant photography or CCD video

FIGURE 5.1: *The structure and application of fluorescent dyes for nucleic acid detection.* In situ *detection of DNA and RNA in both agarose and polyacrylamide gels through the use of fluorescent dyes is the standard tool for detection in molecular biology. The most widely used dye, ethidium bromide, binds nucleic acid reversibly; newly developed dyes such as TOTO-1 and YOYO-1 have a higher affinity allowing samples to be pre-stained before electrophoresis. Fluorescence-stained gels are generally viewed on a transilluminator and photographed with instant Polaroid film, or captured onto a computer format with charge coupled device (CCD) video camera.*

ethidium bromide. As the ethidium bromide releases this energy strong fluorescence is produced. Unfortunately, at a wavelength of 254 nm DNA and RNA suffer considerable molecular damage: breaks, cross-links and base alterations are produced. For this reason it is preferable to irradiate at a longer wavelength. At 302 nm, the ethidium bromide emission of visible red–orange light is somewhat weaker than that at 254 nm, but DNA and RNA damage is drastically reduced. Thus, for most applications ethidium bromide-stained DNA and RNA are irradiated at 302 nm. This can be achieved from above using UV lamps, or from below using a transilluminator (*Figure 5.1*). A transilluminator consists of a box containing UV lamps situated beneath a special glass filter. The filter provides illumination at the critical 302-nm wavelength and a surface onto which gels can be conveniently placed for viewing, photography or cutting out particular bands. Ethidium bromide staining is most commonly used for agarose gels. This is because the yield of red–orange light is reduced in polyacrylamide gels and the gels need to be disassembled before viewing: the glass plates that form the gel sandwich are opaque to UV radiation. Ethidium bromide is capable of detecting in the region of 5 ng of DNA in a single band on an agarose gel. Ethidium bromide-stained gels are best viewed in a darkroom to cut out extraneous sources of light. Examples of ethidium bromide-stained gels and practical details of the methodology can be found in Chapter 7.

It is important to emphasize that even the less damaging 302-nm light can cause significant injury to the skin, and most particularly to the eyes. Always wear a full face shield that has been expressly manufactured for the purpose of excluding UV light. It is a good idea to have two or three shields near the transilluminator so that several members of the laboratory can discuss an illuminated gel at the same time. It is also important to use protection for the hands. Disposable latex gloves are effective at limiting UV exposure. You will of course be wearing your laboratory coat. This gives good protection to the arms, but pay particular attention that you do not expose your wrists when working close to the transilluminator for relatively long periods of time, for example during procedures that call for cutting out and isolating bands.

As might be anticipated for a compound that intercalates into the DNA double helix, **ethidium bromide is known to be mutagenic and is suspected of being a carcinogen.** It is important therefore that laboratories where ethidium bromide is in widespread and routine use treat the compound with respect. Workers should wear gloves when preparing or handling gels or gel tanks that are likely to be contaminated with ethidium bromide. Consult the proper authority in your laboratory regarding the accepted practices for handling such material.

Agarose or polyacrylamide gels that contain ethidium bromide should be collected for incineration with clinical waste. Protocols for inactivating concentrated solutions of ethidium bromide > 0.5 mg ml^{-1} using sodium nitrite and hypophosphorous acid, and for decontaminating dilute solutions with Amberlite XAD-16 or activated charcoal, are described by Sambrook et al. [1]. It was once popular to 'inactivate' ethidium bromide with sodium hypochlorite (bleach). This certainly removes the orange colour of ethidium bromide solutions. The resulting compound is also 1000-fold less mutagenic in the Ames Salmonella/microsome test. However, this procedure is not recommended. Many chemicals only become mutagenic when they are acted upon by the liver (a function performed by the liver microsomes in the Ames test) [2]. Ethidium bromide is one such compound. Surface contamination of the skin is thus not quite so dangerous. However, following treatment with hypochlorite the product of the reaction is now directly mutagenic, without the need for liver activation.

Recently some new fluorescent dyes have been developed by Molecular Probes Inc [3]. These compounds, which have the acronyms TOTO-1 and YOYO-1 to signify the dimerization of their parent compounds TO-PRO-1 (benzothiazolium-4-quinolinolium) and YO-PRO-1 (benzoxazolium-4-quinolinolium), have very high sensitivities of DNA and RNA detection. In addition to their sensitivity, these dyes have the important property of a very low rate of dissociation from nucleic acids. (Ethidium DNA and RNA complexes can be destained by washing.) This permits DNA to be pre-stained with TOTO-1 and YOYO-1 before electrophoresis. As more researchers become aware of these dyes it seems certain that they will be put to some imaginative and productive uses. Whilst these dyes are currently considerably more expensive weight for weight than ethidium bromide, the ability to irreversibly pre-stain fractions of DNA before electrophoresis with very small amounts suggests that they may prove safe and relatively economical to use.

Fluorescence-stained gels are generally recorded by photography. Inexpensive, fixed-focus cameras, originally developed for taking photographs of oscilloscopes, find widespread use in molecular biology laboratories. They use fast (sensitive) ASA 3000 black and white instant Polaroid film (type 667). An orange filter must be used in the camera to cut out all UV light, to which the film would otherwise be sensitive. (Further details of Polaroid photography of ethidium bromide-stained agarose gels can be found in Section 7.10.) Slower, less sensitive instant films are also available which give positive transparencies that can be scanned by a densitometer. Recently, video camera systems have become widely available that are capable of capturing gel images on computer screens and printing the pictures on thermal sensitive paper. Systems such as these can be acquired for the cost of approximately 10 000 Polaroid images. The computer-captured image is of course available for electronic processing, such as densitometry, and storage as a file on computer disks.

5.4 Labeling with radioactive nucleotides

There are three principal radioactive atoms that find widespread use in the detection of nucleic acids following gel electrophoresis: ^{32}P, ^{33}P and ^{35}S. The rationale for the use of the phosphorus isotopes requires no explanation: ^{32}P and ^{33}P can be incorporated in the phosphate backbone or at the 5' terminus of a DNA or RNA strand. The use of the sulfur isotope ^{35}S is surprising. Sulfur atoms are not generally present in nucleic acids. You will recall that sulfur is positioned directly below oxygen in the periodic table of the elements, a placement that indicates similarities in its chemical properties. It has been found that sulfur can replace an oxygen atom in the phosphate component of nucleic acid, without destroying its ability to be incorporated into a nucleotide chain by DNA or RNA polymerase. *Figure 5.2* illustrates the structure of a deoxynucleotide triphosphate (dNTP), in this case dATP, and the positions at which radioactive atoms are commonly placed for applications in molecular biology. Note in particular the designation of the three phosphates as α, β, and γ, counting from the ribose or deoxyribose sugar. Only the α phosphate is incorporated into the polynucleotide chain by DNA or RNA polymerase enzymes. Do not join the legions of students who have made the classic mistake in thinking that the α, β or γ ^{32}P dATP refers to the mode of radioactive decay! ^{32}P emits β-radiation (electrons) not α-particles or γ-rays.

Each of the isotopes ^{32}P, ^{33}P and ^{35}S decays by the emission of a high-energy electron, a β-particle (see *Table 5.1*). ^{32}P is the most energetic, the most sensitively detected and the most dangerous for the research worker. ^{33}P, a relatively recent addition, has a much reduced energy

TABLE 5.1: *Properties and costs of alternative radiolabel for nucleic acids*

Radioactive nucleotide	Energy of β emission (MeV)	Half-life (days)	Maximum specific activities (Ci mmol⁻¹)	Relative cost eg: 9.25 MBq α-dATP
^{32}p	1.71	14.3	6000	1.00
^{33}p	0.249	25.4	3000	1.85
^{35}S	0.167	87.4	1500	1.25

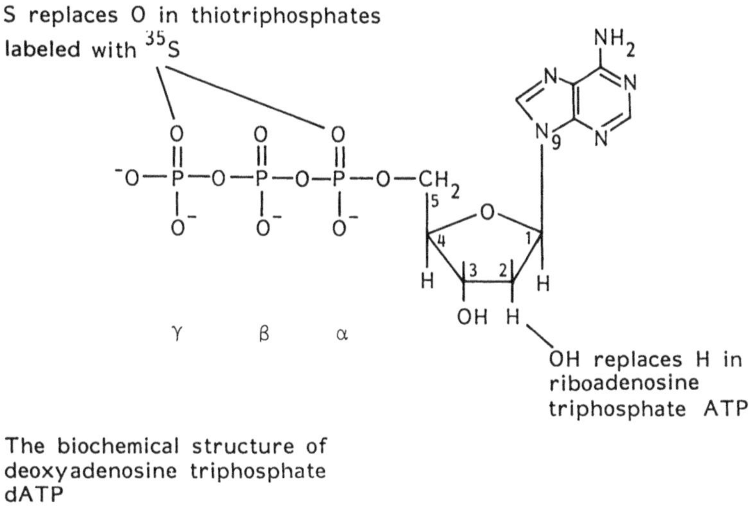

S replaces O in thiotriphosphates
labeled with ^{35}S

The biochemical structure of
deoxyadenosine triphosphate
dATP

OH replaces H in
riboadenosine
triphosphate ATP

FIGURE 5.2: *The positioning of radiolabeled atoms in nucleotides used for nucleic acid detection.*

and is consequently less sensitively detected, but much safer for the operator. ^{35}S has an energy slightly below that of ^{33}P. The half-lives, the time taken for exactly half the radioactivity present at a particular moment in time to decay, are given in *Table 5.1*. The relative costs per unit of radioactivity for an α-labeled deoxynucleotide triphosphate (dNTP) are also indicated. A welcome advance has been the development of stabilization techniques so that solutions of radioactive nucleotides can be stored at 4°C instead of at −20 or −70°C. Inert, but intensely colored dyes are added to the solutions to assist in accurate pipetting and for the identification of the most radioactive tubes during the progress of an experiment.

Safety is of course the primary concern when carrying out any procedure with radioactive material. The high energy of ^{32}P means that steps should be taken to limit exposure. Notoriously, the tips of the fingers and the eyes are the sites that the careless (and the short-sighted) worker irradiates as an unshielded polypropylene microcentrifuge tube is examined to see whether they really did pipette 5 μl of α-^{32}P dATP into the tube. ^{32}P can be used perfectly safely by employing transparent perspex (Plexiglas™) at least 1 cm thick as a screen for operator protection and by using tongs to manipulate tubes containing concentrated solutions of ^{32}P. Work with all radioactive material is governed by local safety rules. Users of ^{32}P should be issued with a film badge, which can be developed to test how much β-radiation a worker has been exposed to within a particular time period. Finger dosimeters, which are worn inside the fingers of laboratory gloves, can also be used. Suppliers of ^{32}P safety equipment are included in the Appendix. ^{33}P and ^{35}S have much weaker β-emissions than ^{32}P and exposure at a distance is less of a problem. Extreme care should still be exercised, particularly with regard to the longer half-lives of these isotopes. Here the danger comes from accidental ingestion, since the radioactive atoms are generally present in molecules that will be readily absorbed and retained by the body. The best protection is to know where the radioactivity is. A hand-held monitor is a prerequisite for safe manipulation of β-radiation. Work areas and pipettes should be monitored both before and after a procedure involving radioactivity.

The most common method for detecting radioactively labeled DNA or RNA is to expose X-ray film to the emitted β-particles: the process of autoradiography. ^{32}P is highly sensitive and rapidly detected, but because of the high energy of decay the resolution of the autoradiographic bands can suffer. ^{33}P and ^{35}S have lower energy and take a little longer to expose, but they produce much sharper autoradiographic images. Gels are generally dried to improve the sensitivity of detection. A recent advance in the detection of radioactive material is the development of a number of phosphorescence image detectors. In these devices special reusable plates are placed over the radioactively labeled gel. The decaying emissions strike the plate and impart an image onto the surface material. Following exposure, the plate is scanned to release the captured information and a high-resolution image is reconstructed by computer analysis of the scanned information.

There are a number of ways in which nucleic acids can be labeled with radioactive atoms. In metabolic labeling cells or tissues are exposed to ^{32}PO$_4$$^{-3}$ or ^{33}PO$_4$$^{-3}$ as phosphoric acid or orthophosphate in solution. A proportion of all the phosphorus atoms in DNA, RNA, lipids and proteins become labeled. More commonly for molecular biological applications, nucleic acids are labeled *in vitro*. These processes are outlined in *Figure 5.3*. Further details can be found in ref. 4. ^{32}P or ^{33}P can be added to 5' ends using the enzyme polynucleotide kinase and γ-labeled ATP. It is only the γ phosphate that is donated to the 5' end of a polynucleotide strand (see *Figure 5.2*). The 3' ends can be labeled with ^{32}P, ^{33}P or ^{35}S using the enzyme terminal transferase and α-labeled dNTPs. Internally labeled DNA can be produced using nick translation, random priming or transcription. During *nick translation* trace amounts of the enzyme DNase I introduce single-stranded nicks into double-stranded DNA. DNA polymerase I is also present and this enzyme binds to the nicks and incorporates α-labeled dNTPs whilst

Solid support

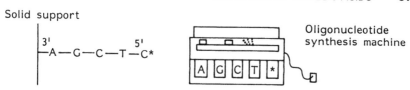

During chemical synthesis of oligo nucleotides (modified nucleotides only)

Terminal transferase

Single base ddNTP

Multiple bases dNTP

3' End labeling

Polynucleotide kinase
(radio label only)

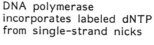

5' End labeling

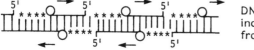

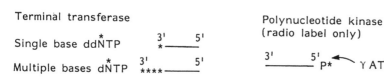

DNA polymerase
incorporates labeled dNTP
from single-strand nicks

Nick translation

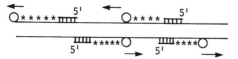

DNA polymerase
incorporates labeled dNTP
from random oligonucleotide
primers

Random priming

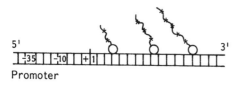

Promoter

RNA polymerase
incorporates labeled
NTPs as it transcribes
a DNA template

in vitro **transcription**

FIGURE 5.3: *Methods for incorporating radioactive or modified nucleotides into nucleic acids. It is important to be clear that you have selected the correctly labeled nucleotide for each of these labeling methods, e.g. α or γ labeled, deoxy or ribo nucleotide. Specific details and protocols for these methods are supplied as part of any proprietary 'kit', or can be found in most collections of molecular biology techniques [4].*

simultaneously degrading unlabeled DNA as it advances in a 5'→3' direction. During labeling with *random primers*, short oligonucleotides of 6–9 bases containing completely random sequences are used to provide random priming sites for DNA polymerase to extend using α-labeled dNTPs. Both nick translation and random priming give rise to labeled DNA of a mixture of lengths and from both strands of a double-stranded template. An alternative method is to use RNA polymerase enzymes, usually from the bacteriophages SP6, T7 or T3, to synthesize RNA *in vitro* from a DNA template using α-labeled NTPs. *Transcriptional* labeling has the advantage of directing the label to one or the other strand of a specific section of DNA and obtaining a fragment with a defined length. Each of the above methods can be employed to label nucleic acids as probes in hybridization experiments. Another important use of radioactive labeling is in the determination of DNA and RNA sequences. Essentially these reactions can be viewed as primer extensions. A polymerase enzyme binds to the 3' end of the oligonucleotide primer that is hybridized to the template. By including an α-labeled dNTP the product is rendered detectable by autoradiography.

5.5 Labeling with fluorescent nucleotides

Techniques that employ labeling of DNA or RNA with fluorescent nucleotides use a fundamentally different principle to detect and record the separation of nucleic acids during gel electrophoresis. Conventional detection methods record the patterns in which nucleic acids are found *after* electrophoresis, either by photography or autoradiography. In the fluorescent nucleotide approach a laser detector scans a fixed position towards the base of the gel and information is gathered *as* the nucleic acids migrate past this position. The impetus for this technology has come from the desire to accelerate and automate the process of DNA sequence determination. The Sanger method of DNA sequencing uses chain-terminating dideoxynucleotide triphosphates to break the chain randomly at every position [5]. The sequence can be read from the ladder of DNA fragments. If a fixed-point detector could automatically distinguish between DNA fragments terminated by dideoxy (dd) ATP, ddCTP, ddGTP and ddTTP, the sequence can be collected into a computer as the DNA fragments migrate down the gel under conditions of denaturing polyacrylamide gel electrophoresis. This eliminates the lengthy and tiresome procedure of making autoradiographs and the manual examination of the sequence information [6].

There are essentially three strategies for fluorescently labeled nucleotide DNA sequencing:

- single fluorescent primer;
- four-color fluorescent primers;
- four-color fluorescent dideoxy terminators.

To understand the difference between these approaches it is important to understand the concepts behind DNA sequencing using the Sanger methodology (see *Figure 5.4*). DNA sequence determination is initiated using a specific oligonucleotide primer that binds adjacent to the region for which sequence information is to be obtained. If a DNA polymerase is added in the presence of the four dNTPs, a molecule complementary to the template strand is made in the 5'→3' direction. ddNTPs are nucleotide analogs in which the 3' sugar hydroxyl is missing, see *Figure 5.2*. When a ddNTP is added to

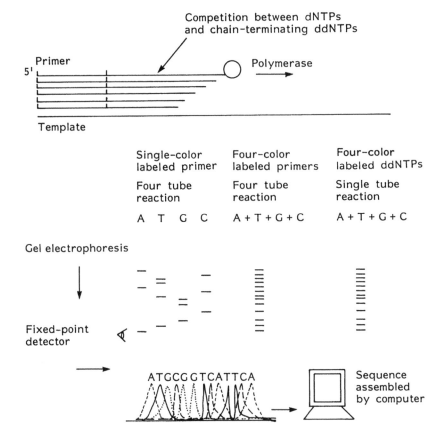

FIGURE 5.4: *Fluorescent nucleotide technology. Fluorescent nucleotides are employed in three different variations of the dideoxynucleotide sequencing methodology developed by Sanger [5].*

the growing nucleotide chain, polymerase extension is terminated and a fragment is produced whose length corresponds to the position of a particular base in the template. By using a separate dideoxy terminator in an optimized ratio with each of dATP, dCTP, dGTP and dTTP, the complete sequence can be determined from the distribution of the various ddNTP-terminated fragment lengths.

In conventional gel electrophoresis and recording technology, four identical reactions are carried out, each with a different ddNTP, and the products are separated in adjacent lanes of a denaturing polyacrylamide gel. Labels that facilitate detection of the DNA can be either radioactive or special molecular tags which direct the binding of specific proteins (see below). These can be placed in the sequencing primer or incorporated during polymerase extension. In the single fluorescent primer method, a fluorescent dye is attached to the 5' end of the sequencing primer and the products of the four ddNTP reactions are loaded as usual in adjacent lanes on the gel. The laser detector knows which terminated fragment corresponds to which template base by the lane in which it is detected. An alternative strategy is to use fluorescent dyes of four different colors. Then the products of the four ddNTP reactions can be combined in a single lane on a gel and the fixed-point detector can read the sequence as each colored fragment passes. There are two approaches to a four-color strategy. On the one hand, four differently colored dyes can be coupled to the sequencing primer (four-color fluorescent primer method). The four ddNTP reactions are carried out separately, but they can then be combined for electrophoresis in a single lane. In the alternative four-color fluorescent dideoxy terminator approach, a differently colored dye is coupled to each of the chain-terminating dideoxy dNTP analogs [6]. Since the fixed-point detector can discriminate between each dye there is no need to carry out these reactions in separate tubes. The one-tube reaction can, of course, be loaded in a single gel track. This approach has the significant added bonus that 'pauses' by the polymerase, which can be mistaken for terminations in other strategies, are eliminated since they are invisible. Fragments can only be detected when they are bona fide ddNTP terminations because the labels are carried by the ddNTPs. The advantage of running all four ddNTP reactions in a single lane is that it removes the problem of deciding the precise order of the fragments, which can arise when the detector has to compare four adjacent lanes on a gel (as in the single fluorescent primer method). However, in the four-color primer approach it has been found that a correction factor has to be applied by the computer software to allow for the effects of the different molecular shapes of the four colored dyes on the rate of migration of the terminated fragments. This does not present a problem in the four-color dideoxy terminator methodology since similar molecular shape dyes are used for each of the four ddNTPs; they each affect mobility to a similar extent. It has been found though that a set of four colored ddNTP terminators has to be optimized for each DNA sequencing enzyme. This is to ensure that each dideoxynucleotide has approximately equal efficiencies of incorporation into the growing polynucleic acid chain. The acceptability of the fluorescently tagged ddNTPs varies from enzyme to enzyme [7].

5.6 Labeling nucleic acids with specific proteins

In the face of the hazards and inconvenience of working with radioactively labeled nucleic acid and, perhaps more importantly, the difficulty of waste disposal, there has been rapid development in recent years of techniques for labeling and detecting DNA nonradioactively. These methods are gaining in popularity and acceptance as they approach the extreme sensitivity of ^{32}P labeling. For the majority of molecular biology applications, where this level of sensitivity is not required, nonradioactive methods are now the protocols of choice. In addition to the lack of hazards, the labeled molecules are extremely stable and can be stored for long periods of time. This is in marked contrast to the inevitable decay of radioactively labeled material. Of course the development of nucleic acid sequencing by fluorescent nucleotides represents a major departure from radioactive technology. However, the drive towards nonradioactive methods is also gathering pace for hybridization-based applications.

Whilst there is some diversity in the details, each of the protein labeling techniques leads to a common end point: the coupling of a DNA or RNA molecule to a sensitively detected enzyme. The most commonly used are horseradish peroxidase (HRP) and alkaline phosphatase (AP). These enzymes can be detected with conventional, colorimetric substrates. However, to a large extent the successful replacement of radioactive methods has derived from the introduction of chemiluminescent substrates for HRP and AP. Chemiluminescence is the production of visible light from a chemical reaction. Light production can be sustained over many hours and an image can be sensitively recorded on X-ray-type films, as in autoradiography, or detected with phosphorescence imaging machines.

The labeling of nucleic acids with specific proteins can be divided into two categories (see *Figure 5.5*):

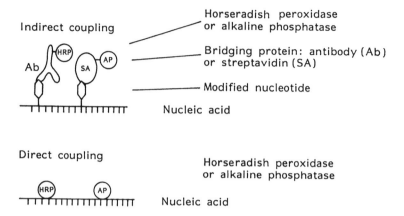

FIGURE 5.5: *Labeling nucleic acids through the use of specific proteins. There has been rapid development in recent years of nonradioactive methods of nucleic acid detection. Most systems use either direct or indirect coupling of horseradish peroxidase or alkaline phosphatase enzymes to generate colorimetric or chemiluminescent signals. Indirect coupling is achieved through the incorporation of modified nucleotides into the DNA or RNA using standard labeling techniques (see Figure 5.3).*

- indirect DNA–protein coupling;
- direct DNA–protein coupling.

5.6.1 Indirect DNA–protein coupling

The indirect labeling of nucleic acid relies on the incorporation of a modified nucleotide into the DNA or RNA. This can be accomplished either during template-directed copying by DNA or RNA polymerases, by 3′ end modification or during the assembly of synthetic oligonucleotides. There are three chemical modifications in common use:

- biotin;
- fluorescein;
- digoxygenin.

Tests have shown that DNA and RNA polymerases and terminal transferase will incorporate nucleotides carrying these groups into polynucleotide chains. Once nucleic acids have been produced containing these molecules, specific proteins are used as a bridge between the nucleotide chain and the ultimate reporter, the HRP or AP enzymes. For biotin-labeled nucleic acids, the bacterial-derived protein streptavidin is used to elicit high-affinity recognition. Binding of the large streptavidin molecule to the biotin 'tag' is facilitated by placing the biotin group at the end of a long spacer chain. Biotin

TABLE 5.2:   *Commercially available direct protein-coupled labeling systems for nucleic acids*

Company	Product	Protein-coupling reagent	Detection
Amersham	ECL direct for nucleic acids	HRP	Chemiluminescent
Biorad	Gene-Lite™	AP-conjugate to oligonucleotides homologous to vector sequences. Probes prepared from vector containing insert	Chemiluminescent
Promega	Lightsmith™	AP	Chemiluminescent

AP, alkaline phosphatase; HRP, horseradish peroxidase.

nucleotides are available with 7, 14 or 21 atom spacers. This helps to eliminate steric hindrance: it allows the streptavidin an unrestricted approach to the biotin ligand. Streptavidin can be coupled or conjugated directly to a reporter enzyme. Alternatively, a biotin-coupled enzyme is added to the streptavidin–biotin–nucleic acid complex. Since streptavidin has many binding sites for biotin, this enables several reporter enzymes to be linked to a single streptavidin–nucleic acid interaction and offers an amplification of the signal. Where nucleotides carry fluorescein or digoxygenin tags, specific monoclonal antibodies, conjugated to either HRP or AP, are used for nucleic acid detection. This can be with either colorimetric or chemiluminescent substrates.

5.6.2 Direct DNA–protein coupling

It is also possible to couple DNA to an HRP or AP enzyme reporter with direct covalent linkages. Of course a nucleic acid that carries a protein moiety will have altered properties in electrophoretic gels. Direct DNA–protein coupling techniques are therefore reserved for applications where nucleic acids are detected after electrophoretic separation through the hybridization of a homologous, direct coupled probe, for example in Southern or Northern blots with which you may already be familiar (see below). Direct DNA–protein coupling has the advantage of being a rapid technique because there is no need for the intermediary steps of nucleotide labeling and streptavidin or antibody detection necessary where indirect techniques are employed. A disadvantage is the limitation that the conditions under which nucleic acid hybridization must proceed are restricted by the need to avoid

TABLE 5.3: Commercially available indirect protein-coupled labeling systems for nucleic acids

Company	Product	Modified nucleotide	Protein-coupling reagent	Detection
Gibco	Photogene™	Biotin-7 dATP Biotin-14 dATP	Streptavidin–AP conjugate	Chemiluminescent
Amersham	ECL for nucleic acids	Fluorescein-dUTP	Anti-fluorescein antibody–HRP	Chemiluminescent
Boehringer	DIG nucleic acid labeling	Digoxygenin-11-dUTP	Fab anti-digoxygenin antibody–AP or HRP conjugate	Chemiluminescent
New England Biolabs	Phototope™	Biotin	Streptavidin followed by AP–biotin conjugate	Chemiluminescent
United States Bio-chemicals	Gene Images	Biotin-21 dUTP	Streptavidin–AP conjugate	Chemiluminescent

AP, alkaline phosphatase; HRP, horseradish peroxidase.

harsh treatments that will destroy the activity of the conjugated enzyme. *Tables 5.2* and *5.3* show some of the commercially available methods that have been developed for direct and indirect protein coupling for purposes of nucleic acid detection. Given the speed of change in this branch of the industry some of these products may have been superseded by the time you read this book. The important point is to convey the breadth of different direct and indirect protein-coupled labeling techniques that have been developed. Further details on labeling nucleic acids with specific proteins can be found in publications from suppliers of molecular biology reagents [8,9].

5.7 Conventional staining

There are a number of conventional staining procedures that find applications for the products of nucleic acid electrophoresis. That in most widespread use is silver staining. Those readers with experience in the field of protein electrophoresis may already be familiar with silver-stain techniques, which can be an order of magnitude more sensitive than Coomassie blue, the most common stain for electrophoretically separated proteins. The advantage of silver staining derives from the convenience of being able to stain nucleic acids directly in a gel without the need for their prior labeling with radioactive or specially modified nucleotides. In this respect silver staining is similar to the use of fluorescent dyes such as ethidium bromide, but unlike ethidium bromide, which performs inefficiently in polyacrylamide gels, silver staining is ideal for the sensitive detection of nucleic acids in polyacrylamide media [10]. However, silver staining is a more lengthy process than ethidium bromide detection: a typical silver-staining procedure may take up to 1 hour. Whilst there are many variations, the principal steps are fixation (usually with acid), impregnation with silver ions and development with Na_2CO_3. Alternative protocols exploit the use of temperature, formaldehyde and thiosulfate to improve contrast and reduce background staining. When correctly performed, silver staining for nucleic acid in polyacrylamide gels gives red/brown to brown/black bands on a colorless background. The sensitivity is at least an order of magnitude above that of ethidium bromide. In contrast to ethidium, silver-stained nucleic acids are not recoverable for further procedures such as cloning or probe preparation.

5.8 Nucleic acid detection *in situ* and after transfer to membranes

Throughout the preceding discussion, only brief reference has been made to whether the products of an electrophoretic separation are to be detected within the gel (*in situ*) or whether they need to be, or would benefit from, being transferred to a support matrix of some kind (a membrane). It is a general requirement that nucleic acids are transferred to the surface of a membrane support in order that they are available for hybridization with a homologous sequence. The classical demonstration of this technique was described by E. Southern in 1975 [11] when it was shown that specific DNA fragments could be detected amongst a large population of electrophoretically separated DNA molecules after they were transferred to a nitrocellulose membrane and hybridized with a radioactively labeled DNA probe. Thus the *Southern blot* was born. *Northern blots* refer to the same process where RNA molecules are transferred from a gel to a membrane. Southern and Northern blots can be probed with DNA or RNA. The classification refers to the nature of the nucleic acid that is transferred to the filter. (*Western blots* are the transfer of proteins from an electrophoresis gel to a membrane. In this instance the probe is usually an antibody or a ligand specific to one or more of the separated proteins.) There do not appear to be any good candidates for an Eastern blot as yet, although some workers have coined the phrase *South-Western blot* for experiments where electrophoretically separated proteins are transferred to a membrane or filter and probed with a nucleic acid to detect DNA or RNA binding proteins [12,13]. Since nucleic acid-binding membranes are composed of porous materials that can be used as filters, the terms 'membrane' and 'filter' are often used interchangeably.

In the original Southern blot, nucleic acids were transferred from an agarose gel on to a nitrocellulose membrane by capillary action (*Figure 5.6*). In this process the gel is placed on to a filter paper that

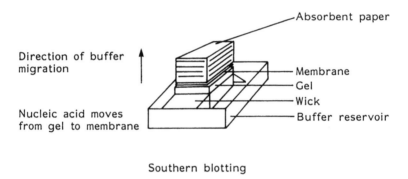

Southern blotting

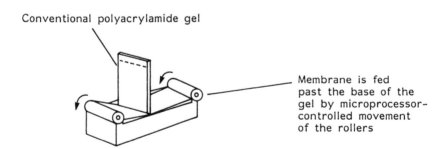

New variations: simultaneous transfer

FIGURE 5.6: *Transferring nucleic acids separated by gel electrophoresis to membranes. The original description of the transfer of nucleic acids separated by gel electrophoresis to a membrane is 20 years old [11]. To permit the more efficient transfer of large gel formats, for example sequencing gels, apparatus capable of simultaneous transfer has been designed. Blotting is accomplished at the same time as nucleic acid separation.*

leads to a reservoir of buffer. A nitrocellulose membrane lies on top of the gel. A stack of dry, absorbent papers, usually paper towels, is positioned on top of the membrane. A heavy weight keeps the arrangement in place. Buffer flows upwards through the gel and into the absorbent papers carrying nucleic acids from the gel and on to the membrane where they are absorbed by noncovalent interactions. In order to make the process more efficient, especially for large molecules, the DNA must first be broken into smaller fragments in the gel. This is accomplished by treating the gel with HCl (acid) and NaOH (base) before assembling the blot.

In the ensuing years, a number of alternative methods have been developed for transferring the nucleic acid out of the gel and on to a membrane.

- Vacuum blotting: a negative pressure is used to accelerate the transfer of nucleic acid from the gel to the membrane.
- Positive pressure blotting: positive pressure is used to force the nucleic acid from the gel to the membrane.
- Electroblotting: an electric field is employed to carry nucleic acid from the gel to the membrane.
- Simultaneous transfer: in these systems (currently applicable to vertical polyacrylamide separations) gel electrophoresis is carried out as usual, but the apparatus is modified so that nucleic acids run out of the base of the gel on to a horizontally positioned membrane (*Figure 5.6*). The membrane is moved past the base of the gel in synchrony with the migration of the nucleic acids, so that the different species are distributed the length of the membrane. By using microprocessors to control the rate at which the membrane moves, the spacing and resolution of the DNA or RNA bands can be carefully controlled.

The nature of the membrane has also evolved in the 20 years since the arrival of Southern blotting. Nitrocellulose membranes can prove to be rather fragile, so many are now reinforced with strong nylon fibers. Alternatively, nylon membranes have been developed with modified surface properties. Original protocols called for membranes to be baked in an oven to fix the absorbed nucleic acid to the membrane sufficiently so that the nucleic acid was retained throughout the considerable number of aqueous washes it received in subsequent hybridization experiments. Many workers now prefer to fix nucleic acids irreversibly to membranes with covalent cross-links. This can be achieved by short wavelength UV irradiation, often with dramatic improvements in the sensitivity of detection. Purpose-built high-intensity UV cross-linking apparatus is available from several manufacturers.

Whilst membrane transfer is essential for experiments involving hybridization detection with labeled probes, there is a trend towards working with membrane-bound transfers of electrophoretically separated nucleic acids for applications where *in situ* detection is possible. This is because membrane-bound material offers a number of distinct advantages. For example, all of the chemiluminescent detection methods require membrane-bound nucleic acid. In addition, applications such as sequencing, footprinting or primer extension require large, thin and consequently fragile denaturing poly-acrylamide gels. Simultaneous transfer to membranes (see *Figure 5.6*) provides a more robust material with which to carry out nucleic acid detection and radioactive detection is more sensitive. Moreover, on a membrane format repeat detections are possible. This facility is exploited in so-called 'multiplex' sequencing. In a multiplex sequence determination, multiple dideoxy sequence determinations can be run in the same lane on a gel. A specific probe can be then hybridized in turn to reveal individual polymerase extension reactions. Large amounts of information can therefore be derived from a single gel by sequential hybridizations with individual probes to the membrane. In this way it is possible to carry out multiple sequence determinations within a single sample of genomic DNA, for example in clinical or forensic applications where a limited amount of material is available.

5.9 Summary of nucleic acid detection methodology

To summarize, there follows a brief guide to the principal methods of nucleic acid detection: their use either *in situ* or after transfer to membranes; their suitability for single- or double-stranded material, and agarose or polyacrylamide gels.

Binding of fluorescent dyes

- Suitable for *in situ* detection.
- Used for agarose and polyacrylamide gels.
- Much more sensitive for double-stranded than single-stranded material; polyacrylamide gels less sensitive than agarose, but abundant PCR material is generally detectable.

Labeling with radioactive nucleotides

- Suitable for *in situ* (better with dried gels) and membrane-bound detection.
- Used for agarose and polyacrylamide gels.

- Used for intrinsic labels or via hybridization with homologous probes.
- Equally sensitive for single-stranded and double-stranded material.

Labeling with fluorescent nucleotides

- Used for *in situ* detection.
- Used predominantly for polyacrylamide gels.
- Used for intrinsically labeled material.
- Detected with fixed-point laser scanners in automated sequencing and PCR fragment analysis.

Labeling nucleic acids with specific proteins

- Only suitable after transfer to membranes.
- Used for agarose and polyacrylamide gels.
- Used for single-stranded and double-stranded material.
- Equally suited for intrinsic labels or via hybridization with a homologous probe.
- Direct and indirect protein coupling available; direct protein coupling restricted to hybridization probes.

Conventional staining

- Used predominantly for detection *in situ* but can be used after transfer to membranes.
- Suitable for double-stranded and single-stranded material.
- Available for polyacrylamide and agarose gels.

References

1. Sambrook, J., Fritsch, E.F. and Maniatis, T. (1989) *Molecular Cloning: A Laboratory Manual,* 2nd Edn. Cold Spring Harbor Laboratory Press, Cold Spring Harbor, pp. 6.16–6.17.
2. Ames, B. (1979) Identifying environmental chemicals causing mutations and cancer. *Science,* **204,** 587–593.
3. Haugland, R.P. (1992) *Handbook of Fluorescent Probes and Research Chemicals,* 5th Edn. Molecular Probes Inc.
4. Sambrook, J., Fritsch, E.F. and Maniatis, T. (1989) *Molecular cloning: A Laboratory Manual,* 2nd Edn. Cold Spring Harbor Laboratory Press, Cold Spring Harbor, pp. 10.2–10.70.
5. Sanger, F., Nicklen, S. and Coulson, A.R. (1977) DNA sequencing with chain terminating inhibitors. *Proc. Natl Acad. Sci. USA,* **74,** 5463–5467.
6. Connell, C. *et al.* Automated DNA sequence analysis. *Biotechniques,* **5,** 342–348.
7. Lee, L.G., Connell, C.R., Woo, S.L., Cheng, R.D., McArdle, B.F., Fuller, C.W., Halloran, D.N. and Wilson, R.K. (1992) DNA sequencing with dye-labeled terminators and T7 DNA polymerase: effects of dyes and dNTPs on incorporation of dye terminators and probability analysis of termination fragments. *Nucleic Acids Res.,* **20,** 2471–2483.

8. Amersham Life Science Catalogue, Amersham International plc.
9. DIG systems users guide for filter hybridisation. Boehringer Mannheim GmbH.
10. Mitchel, L.G., Bodenteich, A. and Merril, C.R. (1994) in *Methods in Molecular Biology,* Vol. 31, Protocols for Gene Analysis, (A.J. Harwood, ed.). Humana Press Inc, Totowa, pp. 197–203.
11. Southern, E.M. (1975) Detection of specific sequences among DNA fragments separated by gel electrophoresis. *J. Mol. Biol.,* **98,** 503–517.
12. Bowen, B., Steinberg, J., Laemmli, U.K. and Weintraub, H. (1980) The detection of DNA-binding proteins by protein blotting. *Nucleic Acids Res.,* **8,** 1–20.
13. Philippe, J. (1994) in *Methods in Molecular Biology,* Vol. 31. Protocols for Gene Analysis, (A.J. Harwood, ed.), Humana Press Inc., Totowa, pp. 349–361.

6 Guide to Techniques and Applications

6.1 Matching the molecule to the technique

Whilst the preceding five chapters have dealt with the basic principles and methods of nucleic acid electrophoresis, in this second part of the book the focus is on the techniques and application of different electrophoretic separations. A consistent format will be used for each chapter. The approach will be to outline the principles of a method and then to focus on published research articles to explain what preparatory steps were taken, how the electrophoretic separation was analyzed and what information was obtained. Recipes and lists of chemicals are not the concern of this introductory text, but for each method suitable sources will be referenced in the Protocols box at the head of each chapter. This will enable a worker with little experience to establish a new technique or refine an existing application once they have become familiar with the fundamental principles described in this book.

In Chapter 1, the various forms in which nucleic acids are found in nature or produced in the laboratory are reviewed (*Figure 1.3*). As a guide to the different techniques and applications within Part 2, *Table 6.1* lists each form of nucleic acid and matches it to the experimental technique in which it is most likely to be encountered, and the chapter in which this application is described. For example, linear single-stranded DNA molecules are encountered in DNA sequencing applications, primer extension reactions, DNA footprinting, SSCP and DSCP. These applications are reviewed in Chapters 10 and 11. It has not been possible to include examples of all of these forms of nucleic acid, but the chapters referred to indicate which type of electro-phoresis would be used in their analysis.

TABLE 6.1: *A guide to the forms of nucleic acids and the electrophoretic methods used in their analysis*

Type of nucleic acid	Size of molecule (bp)	Technique or application	Chapter
Linear double-stranded DNA molecules (1)	10–500	PCR reactions, restriction digests, SSCP, DSCP	7, 10
	500–20 000	PCR reactions, restriction digests, SSCP, DSCP	7, 10
	20 000–2 000 000	Whole chromosomes, genomic mapping	7, 9
Supercoiled, double-stranded circular DNA molecules (2)	2000–50 000	Plasmid analysis, topoisomerase, gyrase assays	7, 10
Nicked double-stranded circular DNA molecules (3)	2000–50 000	Plasmid analysis, topoisomerase, gyrase assays	7, 10
Covalently closed, double-stranded circular DNA molecules (4)	2000–50 000	Plasmid analysis, topoisomerase, gyrase assays	7, 10
Linear single-stranded DNA molecules (5)	10–500	DNA sequencing, primer extension, DNA footprinting, SSCP, DSCP	10, 11
	500–20 000	Viral genomes, *in vitro* mutagenesis, single-stranded probes	8
Circular single-stranded DNA molecules (6)	2000–50 000	Viral genomes, *in vitro* mutagenesis, single-stranded probes	8
Interlocked double-stranded DNA circles (7)	2000–50 000	Topoisomerase, gyrase assays	7, 10
Single-stranded RNA circles (8)	2000–50 000	Testing mechanism of initiation of protein synthesis	8, 11
Double-stranded linear RNA molecules (9)	10–500	Analysis of viral genomes	10
	500–20 000	Analysis of viral genomes	7
Single-stranded linear RNA molecules (10)	10–500	mRNA, tRNA, rRNA analysis and purification, Northern blotting	10, 11
	500–10 000	mRNA, tRNA, rRNA analysis and purification, Northern blotting	8, 11
Branched single-stranded RNA molecules (11)	10–500	Mechanism of intron–exon splicing	10, 11
	500–10 000	Mechanism of intron–exon splicing	8, 11
RNA:DNA hybrid molecules (12)	10–500	S1 mapping, nuclease protection assays	11

DSCP, double-stranded conformational polymorphism; PCR, polymerase chain reaction; SSCP, single-stranded conformational polymorphism.
Numbers in parentheses refer to structures illustrated in *Figure 1.3*.

6.2 Strategic considerations: avoiding elementary mistakes

There are a few universal golden rules that apply to any form of electrophoresis. If they are adhered to, many of the more common elementary mistakes and misapprehensions will be avoided. These rules will be part of the culture in any laboratory with substantial experience and success using nucleic acid electrophoresis. It is particularly important that you adopt them as part of your laboratory culture.

(1) In what form is the nucleic acid? Is it double stranded or single stranded? Are both strands or only a single strand labeled? Is it necessary to use denaturing conditions to separate strands and remove intramolecular base pairing, or do you need native conditions where double-stranded molecules will remain base paired?

(2) What size molecules will you be dealing with? It is a common mistake to run small-sized molecules off the end of a gel because larger molecules are 'expected'. On the other hand the molecules of interest may have completely failed to enter the gel.

(3) The importance of including the appropriate positive and negative controls on each gel cannot be overemphasized. The two most important controls in electrophoretic separations are the inclusion of size markers and the inclusion of a positive and negative control for the detection process. Often, molecular size markers can fulfil both roles. This point will be constantly reiterated in the chapters that follow.

7 Nondenaturing Agarose Gel Electrophoresis

Applications

Double-stranded DNA molecules: 500–20 000 bases.
- PCR amplifications.
- Restriction digests.
- Research application: Southern Blotting (see Section 7.13).

Protocols

Molecular Cloning: A Laboratory Manual, Chap. 6, Gel electrophoresis of DNA, pp. 3–35 [1].

Molecular Biology LabFax, Chap. 8, Electrophoresis, pp. 255–261 [2].

FMC BioProducts Catalog. Technical Applications, pp. 68–93 [3].

Nondenaturing agarose gels are the most widespread and versatile gel electrophoretic techniques for the analysis of nucleic acids. These gels are used when double-stranded molecules are present. For DNA this is the most commonly encountered format. Double-stranded DNA molecules are generated in PCR reactions and in restriction enzyme digests, which are used for the monitoring of cloning and subcloning procedures that form the core of much of molecular biological research and analysis. Nondenaturing agarose gels are therefore indispensable for this discipline.

As nondenaturing agarose gels are so commonly employed, technically straightforward and very much the key technique for a beginner in the field of molecular biology, I have chosen to describe this technique in some detail. The effects of different field strengths, different buffers and agarose gel concentrations are illustrated with examples. At the end of the chapter a research application is reviewed. In subsequent chapters the need to keep this basic introductory text to a manageable size dictates that other forms of nucleic acid electrophoresis are described solely through an explanation of a research application. However, the strategies outlined in this chapter apply in a general sense to most electrophoretic techniques. The reader who identifies an application from subsequent chapters will be able to consider the variations employed in this chapter, and use them as they perfect a technique to suit their own needs.

7.1 Buffers for nondenaturing agarose gels

The role of buffer ions in gel electrophoresis was described in Section 2.1. There are two buffers in common use: Tris-acetate-EDTA (TAE) and Tris-borate-EDTA (TBE). Consult the references in the Protocols box at the head of this and subsequent chapters for the exact formulations and technical details. TBE is often stated to provide improved resolution. A comparison of each buffer system is shown in *Figure 7.1*. Each gel was loaded with mixtures of double-stranded linear DNA fragments generated by *Hin*dIII and *Hin*dIII–*Eco*RI restriction enzyme digestion of bacteriophage λ DNA. The gels were run under otherwise identical conditions. After electrophoresis the gels were stained with ethidium bromide and photographed on a transilluminator (see Section 5.3). The length of the DNA fragments is listed to the side of each gel. The exact size of every fragment is known since the complete nucleotide sequence of the bacteriophage was established in 1982 [4]. Examination of *Figure 7.1* shows that at least under these conditions there is little to choose between the separative powers of TBE over TAE buffer. In TAE gels migration is a little faster and consequently the fragments have moved further down the gel. Inspection of pairs of closely spaced bands reveals that the discrimination is of comparable power in each buffer for the majority of fragment lengths in the mixtures of *Hin*dIII and *Hin*dIII–*Eco*RI digested λ DNA. Some post-separation procedures such as extraction with glass–milk matrices (e.g. Geneclean) work better with TAE gels than TBE (although modified protocols remove this qualification). One can therefore switch between the two buffer systems with confidence that the separations in one buffer will be reproduced in the other. However, when electrophoresis is carried out for extended periods, the buffer can become depleted by the generation of H^+ and

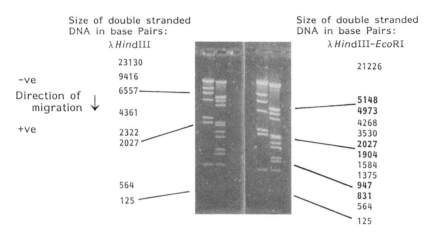

TBE   TAE

Size of double stranded
DNA in base Pairs:
λ *Hind*III

Size of double stranded
DNA in base Pairs:
λ *Hind*III–*Eco*RI

-ve

Direction of
migration   ↓

+ve

23130
9416
6557
4361
2322
2027
564
125

21226
5148
4973
4268
3530
2027
1904
1584
1375
947
831
564
125

FIGURE 7.1: *A comparison between Tris-acetate-EDTA and Tris-borate-EDTA buffers for nondenaturing agarose gel electrophoresis of linear double-stranded DNA fragments. Gels contained 0.8% agarose in buffer and were subjected to a field of 5 V cm⁻¹ (60 V) for approximately 3 h by which time the bromophenol blue tracking dye had reached three-quarters the length of the gel. Following electrophoresis each gel was stained with 0.5 µg ml⁻¹ ethidium bromide in H₂O for 30 min and destained for 5 min in H₂O. Each lane was loaded with 750 ng DNA. The DNA samples were derived by restriction enzyme digestion of bacteriophage λ DNA, a linear double-stranded DNA molecule of 48 502 bp. λ was cut with the enzyme* Hind*III or both* Hind*III and* Eco*RI. The* Hind*III digest is on the left and the* Hind*III–*Eco*RI is on the right of each gel. The fragment sizes are to the left and right of the figure respectively. The wide range of fragment sizes generated by these enzymes make these two λ digests popular as a source of size markers for agarose gel electrophoresis. Never run a gel without some form of marker DNA.*

OH⁻ at the electrodes (see Sections 2.1 and 4.1). TBE has a greater buffering power than TAE, so it is less prone to depletion.

7.2 The resolving power of agarose: altering the gel concentration; altering the field strength

The gels in *Figure 7.1* contained 0.8% agarose. Fragments ranging from about 20 000 bp to 500 bp are spread evenly down the length of the gel. The resolving power of the matrix, the different sized molecules that the gel can discriminate, is illustrated by examining three pairs of fragments in the *Hind*III–*Eco*RI digested λ DNA which are identified in *Figure 7.1*.

5148–4973	175 bp = 3.4%	barely resolved
2027–1904	123 bp = 6.07%	resolved
947–831	116 bp = 15.5%	well resolved

Whilst each of the three pairs of fragments differ by somewhat similar quantities, they are resolved to different extents. Clearly it is the difference in molecular length as a percentage of the total fragment length that is the important parameter. The separative properties of the agarose matrix derive from its ability to retard the migration of end-on reptating linear fragments as described in Chapter 2. The effect of changing the agarose gel concentration to give smaller or larger pore sizes is shown in *Figure 7.2*. At lower concentrations of agarose, the DNA molecules reptate more quickly through the gel and migrate further in a given length of time. The most pronounced differences occur for the fragments of 6557 bp and above. These are very well separated in 0.4% gels, adequately resolved in 0.8%, but completely unseparated in the 1.6% gel. In conclusion, if there are bands of interest between 5000 and 20 000 bp, an agarose gel of 0.4% would be advised. If bands below 2000 bp are of interest, a 0.4% gel would be inappropriate, but a 1.6% gel would give good resolving power. The 1.6% gel shows good resolving power between the 564-bp fragment and the barely visible 125-bp band. A reasonable compromise for the range of fragment lengths in the *Hind*III and *Hind*III–*Eco*RI λ digests is the 0.8% gel. This gives acceptable resolution at both the large and the small end of the range. In practice, gels of around 0.8% agarose are run routinely, with larger or smaller pore gels being selected for particular ranges of fragment lengths. High-percentage agarose gels (2–4%) are used extensively for analyzing the products of PCR amplifications (see Section 4.4 for a discussion of modified agarose preparations that are available to improve the performance of high-percentage agarose gels).

Notice that the shorter the length of DNA fragment, the less intense the staining with ethidium bromide. Ethidium bromide binds multiple times along the length of the molecule. In a restriction enzyme digest that has gone to completion each fragment is present in equimolar amounts, that is there are the same number of molecules of each fragment. Larger fragments have more binding sites for ethidium bromide so they stain brighter. In a restriction enzyme digest the intensity of ethidium bromide staining decreases from the top to the base of the gel. A band that stands out brighter than the band above it may well be a mixture of several fragments running together. In contrast, radiolabeling techniques that incorporate at the

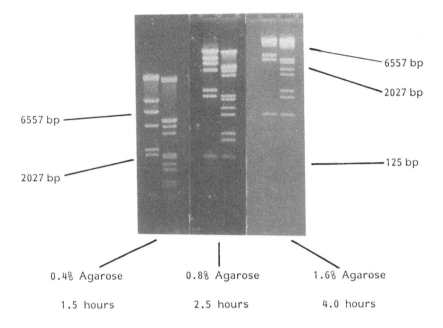

6557 bp

2027 bp

6557 bp

2027 bp

125 bp

0.4% Agarose   0.8% Agarose   1.6% Agarose

1.5 hours   2.5 hours   4.0 hours

FIGURE 7.2: *A comparison of the effect of gels of different agarose concentrations on nondenaturing gel electrophoresis of linear double-stranded DNA fragments. Gels were prepared with 0.4, 0.8 and 1.6% agarose in TBE buffer and loaded with λ HindIII and λ HindIII–EcoRI marker DNA as in* Figure 7.1. *Electrophoresis was at 5 V cm⁻¹. After the given times the gels were stained in 0.5 µg ml⁻¹ ethidium bromide as in the legend of* Figure 7.1.

ends of the molecule detect all fragments with equal intensity. To see the 125-bp band clearly, more DNA would need to be loaded but then the larger bands would be overloaded (see Section 7.7).

Another important consideration is the time during which migration is permitted. It can be seen from *Figure 7.2* that the higher the percentage of agarose, the longer electrophoresis needs to be maintained. Of course, the time taken to complete a separation can be reduced, by increasing the field strength or the constant voltage applied. However, turning up the voltage is a step that should be exercised with caution. *Figure 7.3* illustrates the effect of raising the field strength on the separation of the mixtures of *Hind*III and *Hind*III–*Eco*RI digested λ DNA. The 0.8% agarose TBE gels in *Figures 7.1* and *7.2* were run at 5 V cm⁻¹ (60 V on this apparatus) and took between 2.5 and 3 h to complete. Increasing the field strength to 12.5 V cm⁻¹ gives an acceptable separation at the expense of some slight bunching of the bands into a smaller region of the gel and increased smearing of the larger fragments. However, the run was completed within 48 min, a considerable time saving. Further increases in field strength are not recommended. At 18.3 V cm⁻¹ smearing and bunching are increased further still (30-min separation). At 27 V cm⁻¹, although the run took a mere 14 min, smearing is too great to allow convincing identification of individual bands and the fragments are bunched into two-thirds the length of gel they occupy at 5 V cm⁻¹. The reason for this bunching of fragments at higher field strengths is explained in detail in section 2.2. Briefly, as field strength increases larger fragments of DNA adopt a rigid rod conformation in place of the flexible reptating molecules found at lower voltages. This destroys the dependence of length on migration, whilst increasing the overall velocity: the separative power of the matrix is lost. To maintain a high quality in your electrophoretic separations I would caution against exceeding 10 V cm⁻¹. On the other hand, it is possible to set the voltage too low. At moderate agarose concentrations and low field strengths (2.5 V cm⁻¹) a phenomenon

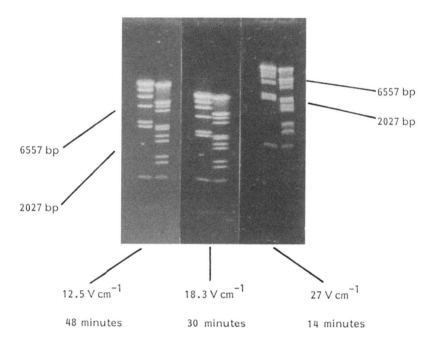

FIGURE 7.3: *The effect of increasing field strength on separation and run time during nondenaturing agarose gel electrophoresis of linear double-stranded DNA fragments. Gels were prepared with 0.8% agarose in TBE buffer and loaded with λ HindIII and λ HindIII–EcoRI marker DNA as in* Figure 7.1. *Electrophoresis was carried out at 12.5, 18.3 and 27 V cm⁻¹ (150, 220 and 320 V for the 12-cm electrode separation in this particular gel aparatus). After the given times the gels were stained in 0.5 µg ml⁻¹ ethidium bromide as in* Figure 7.1.

known as band inversion occurs (see Section 2.2). Fragments can get trapped in the matrix and all certainty that a band at a position on the gel represents a shorter species than the one above it is thrown into question. This is perhaps the best reason for always running an easily recognizable mixture of marker DNA fragments on every gel. The potential for band inversion is also a good reason to avoid long slow overnight electrophoresis that is still running the next morning. It is probably better to set the voltage at an intermediate level using a power pack timing device. Unless very small fragments of the order of 200 bp or less are being separated (and then you would be advised to use a polyacrylamide gel), there is no detectable diffusion of bands from the gel in the absence of an electric field (see Section 7.6).

7.3 Estimating the length of unknown fragments using semi-log plots

Frequently electrophoretic separations generate PCR products or restriction enzyme fragments whose lengths need to be estimated with some precision. Of course this is only possible if some markers of known size are run in adjacent lanes on the same gel. Accurate estimates of size can be obtained by plotting the dependence of length against the relative migration. This is illustrated in *Figure 7.4*. The base 10 logarithm of the molecular length gives a straight-line relationship when plotted against the relative migration for all but the longest fragments: larger fragments move faster than expected (see *Figure 7.4*). Here is a step-by-step guide to performing a semi-log plot to estimate the size of an unknown linear DNA molecule.

(1) Using a photograph or autoradiogram of the gel measure the distance from the wells to the furthest point reached by the tracking dye or measure the distance from the wells to the end of the gel.

(2) Measure the distance from the wells to a common point within each band, i.e. the leading edge, the trailing edge or the mid-point of a band. Whichever you choose, be consistent and measure to within 0.5 mm.

(3) Express the migration of each band in relation to the total distance in (1) above: for example 60/110 mm gives a relative migration of 0.60.

(4) For each of the markers plot the length in base pairs against the relative migration value from (3) above on a base 10 logarithmic scale, i.e. 1000 bp = 3.00, 10 000 bp = 4.00, etc. Semi-logarithmic graph paper is the most straightforward method (see *Figure 7.4*).

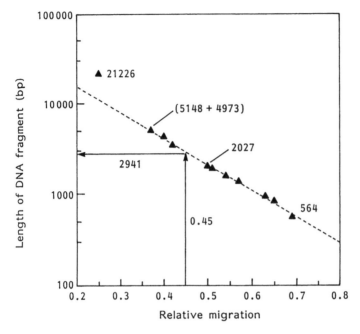

FIGURE 7.4: *Semi-logarithmic plots of relative migration vs. DNA fragment length can be used to obtain an accurate estimate of the length of an unknown molecule. The migration of the λ HindIII–EcoRI fragments in the 0.8% gel run at 5 V cm⁻¹ in Figure 7.1 were used to construct this graph. The length of an imaginary fragment having a relative migration of 0.45 is calculated to represent a length of 2941 bp. Details of the calculation can be found in Section 7.3.*

(5) Draw a straight line through the points. Only include consecutive points that will fall on a straight line and include the region containing the unknown fragments whose length you wish to measure. (In *Figure 7.4* the largest band clearly does not fall on the straight line and has not been included in the plot.)

(6) The molecular length of the unknown fragment is discovered by connecting a perpendicular line from the relative migration axis to the straight-line plot and reading the value on the molecular length axis.

(7) If you have a calculator or computer graph-plotting program, carry out a linear regression to find the best fit line using the calculated slope (m) and intercept (c) of the equation for a straight line ($y = mx + c$) to find the unknown molecular length precisely. For example in *Figure 7.4* the computer gives the slope as –2.87 and the intercept as 4.76. Substituting these in the equation: y (the unknown length) = –2.87 x 0.45 (the relative migration from the x axis) + 4.76 (the intercept) = 3.47. The anti-log of 3.47 is 2941. This is the length in base pairs of the unknown fragment.

(8) Always check that your calculations make sense! Looking back at the TBE gel in *Figure 7.1* (the gel from which *Figure 7.4* was constructed), an imaginary band midway between the 2027 and the 3530 λ *Hin*dIII–*Eco*RI markers would be expected to have a size of something of the order of 2900 bp. You can trust your arithmetic!

7.4 Solving a separation problem: resolving the 5148 and 4973 bp bands in the *Hin*dIII–*Eco*RI restriction digest of phage λ DNA

Imagine you were carrying out a procedure to distinguish between two DNA fragments comparable to the 5148 and 4973 bp fragments in the *Hin*dIII–*Eco*RI digest of phage λ DNA (see *Figure 7.1*). They might represent alternative products that would allow you to identify the orientation of a cDNA fragment cloned into a plasmid. Expression of the cDNA will only occur in the correct orientation. Alternatively, a human genetic disease may be linked with a sequence variation that generates a restriction fragment length polymorphism (RFLP). In individuals with the disease the presence of the site gives a 4973-bp band. In individuals without the disease the position of the site gives a 5148-bp band. How would you proceed in designing agarose gel conditions that successfully distinguish these two products?

There are two variables that control the resolution of an electrophoretic separation which we have reviewed so far in this chapter:

* agarose concentration;
* field strength.

To these we should add two more variables:

* length of the gel: the greater the distance that DNA fragments have to travel, the more likely it is that they will become separated. However, as the fragments travel further, the individual bands get broader (an illustration of the range of different sized gels that are available can be found in *Figure 4.2*);
* time of electrophoresis: the length of time that electrophoresis is carried out will also determine, in any one set of conditions of agarose concentration and field strength, whether two molecules can be resolved.

Inspection of *Figure 7.1* shows a hint of separation of the 5148 and 4973 bp bands at 5 V cm^{-1} in 0.8% agarose. Electrophoresis had been carried out for 3 h. Perhaps 1% agarose would offer a chance of resolving these two fragments. Since higher agarose concentrations increase the time taken to migrate, the field strength could be increased to 7.5 V cm^{-1} without suffering a loss in resolution (see *Figure 7.2*). A good guide to the progress of electrophoresis is to monitor the movement of tracking dyes that migrate at the same rate as different sized DNA molecules. *Figure 7.5* shows the course of a separation between the 5148 and 4973 bp bands in a 1% agarose gel at 7.5 V cm^{-1}. The tracking dyes in this case are xylene cyanole and bromophenol blue. These blue dyes are visible without staining throughout electrophoresis. They migrate at approximately the same rate as DNA fragments of about 9000 and 500 bp respectively. *Figure 7.5* shows the gel at hourly intervals. At 1 h, four molecules (5148, 4973, 4268, 3530) are bunched into just two bands. After 2 h, three bands are visible, but the 5148 and 4973 fragments are not resolved. After 3 h, the critical bands are separated. A further period of electrophoresis does not improve the resolution. After 4 h the fragments are no more clearly defined than they were after 3 h. Variation of these same parameters (matrix concentration, field strength, length of gel and time of electrophoresis) can all be applied to most other techniques of gel electrophoresis discussed in subsequent chapters.

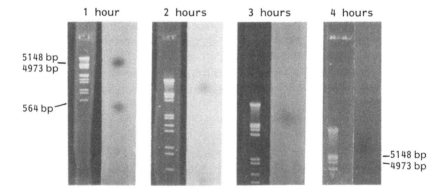

FIGURE 7.5: *Resolution between 5148 and 4973 bp fragments of double-stranded DNA by extended electrophoresis. Nondenaturing TBE electrophoresis was carried out in 1% agarose with a field strength of 7.5 V cm^{-1} (95 V). The left-hand gel in each group shows the pattern of λ HindIII–EcoRI DNA fragments following ethidium bromide staining. The right-hand gel shows the migration of xylene cyanole (upper marker) and bromophenol blue (lower marker) dyes present in the loading buffer. The gels show the migration of both DNA and marker dyes during 4 h of electrophoresis.*

7.5 Alternative methods for staining with ethidium bromide

The example in the previous section demonstrated the use of visible tracking dyes to monitor the progress of an electrophoretic gel. Up to this point, all the examples of agarose gels have been stained after electrophoresis by soaking the gel in H_2O containing 0.5 μg ml^{-1} ethidium bromide followed by a brief period of destaining. However it is possible to carry out electrophoresis in the presence of ethidium bromide and in this way periodically remove the gel to the surface of a transilluminator to monitor the migration of the bands of interest. This is obviously much easier if the gel is held in a tray that is transparent to UV light (see Section 4.1). *Figure 7.6* illustrates the appearance of gels stained in three different ways.

(1) Ethidium bromide was present in the running buffer. Note that the positively charged ethidium molecule migrates as a dye front up the gel in the opposite direction to the negatively charged DNA.
(2) Ethidium bromide was present in the agarose. Again, the positively charged ethidium molecule migrates as a dye front up the gel in the opposite direction to the negatively charged DNA.
(3) Ethidium bromide was present in the loading buffer. In this case there is no ethidium dye front moving in the gel. Ethidium bromide is bound to the DNA molecules and migrates with the fragments.

As ethidium bromide carries a positive charge, it will neutralize a portion of the negative charges present on the DNA molecule. Consequently, the migration of DNA towards the positive electrode is reduced relative to DNA in the absence of bound ethidium bromide. Of more significance is the effect that intercalation of the planar ethidium bromide molecule has in reducing the flexibility of the DNA fragment. This restricts the ability of the fragments to reptate through the pores of the agarose matrix. This is most readily seen in *Figure 7.6c* where the DNA is retarded in proportion to the quantity of ethidium bromide present in the loading buffer. It is essential therefore that the migration of DNA fragments is always compared in the same conditions or concentration of ethidium bromide.

In conclusion, the best resolution and performance with agarose gels is obtained by staining after electrophoresis. This can be somewhat inconvenient in routine molecular biological procedures. Often it is decided to run a gel a little longer to try to separate some closely

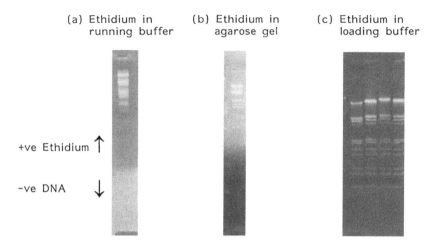

FIGURE 7.6: *Alternative methods for staining nondenaturing agarose gels with ethidium bromide. Gel electrophoresis can be carried out in the presence of ethidium bromide. When placed in either (a) the running buffer, (b) the agarose gel or (c) the loading buffer, DNA samples can be stained and their migration monitored periodically during electrophoresis. Ethidium bromide is a positively charged molecule (see section 5.3) and it migrates in the opposite direction to the DNA. When ethidium bromide is in the running buffer (a) or in the gel itself (b), an ethidium bromide dye front moves from the bottom to the top of the gel. No dye front is seen when ethidium bromide is mixed with the DNA for a few minutes in the loading buffer (c). Lane 1 contained 0 ng ethidium bromide mixed with 750 ng DNA; lane 2 contained 20 ng ethidium bromide mixed with 750 ng DNA; lane 3 contained 2000 ng ethidium bromide mixed with 750 ng DNA; lane 4 contained 200 ng ethidium bromide mixed with 750 ng DNA. The retardant effect of ethidium bromide on the electrophoretic migration of DNA is clearly seen. Lane 1 becomes stained when ethidium bromide enters the running buffer and diffuses into the gel.*

spaced bands. Often the gel may simply be a device to check that one enzymatic step, a restriction or ligation reaction in a multistep process, has proceeded as expected. In these circumstances the addition of some ethidium bromide to the running buffer, the agarose gel or the loading buffer will stain the DNA whilst electrophoresis is in progress and avoid the need to stain the gel. Although ethidium bromide will gradually elute from DNA, even in conditions where the gel is left for extended periods of time (overnight) in ethidium bromide-free buffer, there is no detectable reduction in staining intensity (see Section 7.6 and *Figure 7.7*). Remember, if it is desired to resume a run where DNA has already been stained and destained following an initial period of electrophoresis, it is essential that the staining and destaining is performed in running buffer so that the ionic conditions in the gel do not differ from those in the electrophoresis apparatus.

7.6 Passive diffusion: can agarose gels be left safely overnight before photography?

Often it is not possible to be present the instant an electrophoretic separation is completed. The use of power packs with integral timers or the use of domestic plug-in timing devices that cut the power to the gel are highly convenient ways to organize laboratory time. An important question to answer then is: Do DNA fragments diffuse out of gels in the absence of an electric field? If this was the case an otherwise sharp separation might appear blurred or smeared. *Figure 7.7* shows the result of leaving an agarose gel that had been stained by including ethidium bromide in the loading buffer to remain overnight in the electrophoresis tank. There is no evidence for diffusion out of, or within, the gel, even for the short molecular length fragments.

7.7 Amounts of DNA: estimating unknown quantities and overloading or under-loading the gel

7.7.1 Estimating unknown quantities

Whilst there are more accurate and sophisticated ways for estimating the amount of DNA in a sample of plasmid or viral DNA (e.g. UV absorption at 260 nm or fluorimetry with the DNA-specific dye Hoechst 33285), a rough and ready method is to compare the intensity of ethidium bromide-stained bands on a gel. Ethidium bromide stains in proportion to the length of DNA present. In a restriction enzyme digest, each fragment is present in equivalent molar amounts. There are the same number of molecules in each band, but the longer the fragment the greater the mass of DNA. Given an unknown sample, the quantity of DNA in a band can be estimated by comparing it to a marker fragment with similar intensity. For example, the 2027-bp fragment of a λ *Hin*dIII–*Eco*RI digest is 2027 bp out of a total of 48 502 bp in the complete genome. In a sample of 750 ng, the 2027-bp band

Photographed
immediately

Photographed after
overnight storage
in buffer

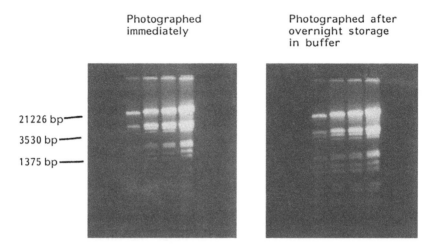

21226 bp

3530 bp

1375 bp

FIGURE 7.7: *Passive diffusion: can agarose gels be left safely in running buffer overnight before photography? The two images represent the position of bands in the gel pictured in* Figure 7.8 *after 1 h of electrophoresis: (a) was taken immediately, whilst (b) was taken 10 h later, during which time the gel was stored in the gel tank immersed in TBE running buffer at room temperature. The only detectable difference is that the stored gel has swollen in size to a small degree.*

contains 2027/48 502 x 750 ng = 31.3 ng DNA. When comparing the staining of bands in this way it is important to load several dilutions of the unknown sample and not to stain the gel with too much ethidium bromide as this can overemphasize the lower molecular weight bands. In fact, adding a little ethidium bromide to the loading buffer probably gives the most accurate assessment. Even so, the relationship between intensity and quantity is linear only over a restricted range. Video cameras are used increasingly to capture images of ethidium bromide-stained gels (see Section 5.3). An electronic picture is readily analyzed by computer image analysis programs, of which a large number are available. Whatever the complexity of the software or video camera system, there is no escaping the fact that reliable estimates can only be made where a linear relationship between band intensity and the quantity of known standards has been established on the same gel and that the intensity of your unknown sample falls within this linear range.

7.7.2 Overloading and underloading a gel

Clearly, there must be a limit to the capacity of the agarose matrix to accommodate migrating DNA molecules at any particular point. The upper limit of the amount of DNA that can be loaded depends on its distribution amongst a few or a large number of different sized molecules. Where a large number of fragments are present at many sizes, for instance a sample of mammalian DNA digested with *Eco*RI, the capacity of a modestly sized gel is of the order of perhaps 5–10 µg. Where most of a sample is present in a single fragment of discrete length, much less DNA can be loaded. In this case all the material will try to occupy the same physical space within the matrix. When the matrix capacity is overloaded, smeared and distorted bands result. *Figure 7.8* shows the result of loading 325, 750, 1500 and 3000 ng of DNA as a λ *Hind*III–*Eco*RI digest. The largest molecular weight bands are overloaded in the 1500 and 3000 ng samples. They run as smears making it difficult to determine their relative migration on the gel. The 21-kb band in the 750 ng sample gives acceptable resolution. This band represents 21/48 kb = 0.44 × 750 ng. Thus the maximum load for a single band in this gel is about 330 ng. The quantity of DNA that can be run before smearing depends on the thickness of the gel, ultimately on the dimensions of the well. A very thin well will crowd the DNA into a narrow region, whereas a thick well will spread it through a greater volume of matrix. There is of course a lower limit below which quantity ethidium bromide-stained DNA will not be visible. The 564-bp band from the λ *Hind*III or *Hind*III–*Eco*RI digest

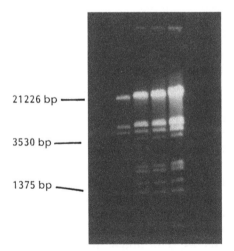

FIGURE 7.8: *Detecting different amounts of DNA by ethidium bromide staining. The gel lanes were loaded with a total of 35, 750, 1500 and 3000 ng of λ HindIII–EcoR1 fragments. Ethidium bromide (200 ng) was added to the DNA and loading buffer mix just before the samples were introduced into the wells. The gel (0.8% agarose) was run at 7.5 V cm^{-1} for 2 h. After the first hour the gel was left for 10 h in the gel tank submerged in running buffer.*

is easily detectable in gels that have been stained after electrophoresis (see *Figures 7.1* and *7.2* for examples): 564 bases represents 0.564/48.5 kb × 750 ng = 8.7 ng of λ DNA. On this reckoning 5 ng would be approaching the lower limit of detection in a single band for gels of this dimension (50 ml agarose as a 10 cm × 7.5 cm × 0.75 cm slab).

7.8 DNA conformation and the mobility of molecules in agarose gels

Molecular biology is frequently conducted with circular double-stranded DNA molecules in the form of bacterial plasmids. A double-stranded DNA circular plasmid can be converted into a linear fragment with a single restriction enzyme cut. It is important to realize that a circular or linear conformation, although there is the same mass of DNA, can have profound effects on mobility in an agarose matrix. In the circular conformation, molecules can exist as either a supercoiled or an open circular form (see *Figure 1.3*). Supercoiling describes the twists put into a DNA strand that results in the molecule twisting back on itself and adopting a more compact conformation. By convention, negative supercoils are those that tend to open up a DNA double strand. Positive supercoils describe twisting in the opposite direction. In contrast, open circular forms contain no supercoils and represent a molecule with a larger radius of gyration (see Section 2.2). The split between the amount of a plasmid preparation that is present as supercoiled or open circular plasmids is determined by the conditions under which the plasmids were isolated. A single-strand nick is sufficient to convert supercoiled plasmid into open circular material, and indeed most circular DNA that is not supercoiled will be present as nicked circular molecules. Examples of nicked circular, supercoiled and linear molecules run on an agarose gel are shown in *Figure 7.9a*. These molecules are sometimes known as form 1 (supercoiled), form 2 (nicked circles) and form 3 (linear) species. Lane 2 shows a preparation of a 7.1-kb plasmid. The lower band contains the supercoiled molecules. The upper band contains nicked, open circular species. The linear plasmid, cut with a restriction enzyme for which there is a single site, is shown in lane 3. An accurate size can only ever be determined on the basis of the migration of a linear fragment compared to a set of linear markers. Even trying to compare the relative sizes of different plasmids where the DNA is circular can prove to be very unreliable. Gel conditions, such as voltage, agarose concentration or ionic strength, can alter the relative mobilities of circular and linear DNA molecules. Moreover,

(a) No ethidium   (b) With ethidium

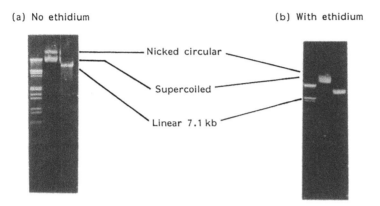

FIGURE 7.9: *Plasmid conformation and migration in agarose gels. (a) Lane 1 contains λ HindIII–EcoRI markers, lane 2 contains approximately 800 ng of uncut plasmid. The upper band is nicked circular DNA, the lower band contains supercoiled material. Lane 3 contains the same amount of material as a single linear fragment following restriction enzyme digestion. The samples were run on a 0.8% agarose gel at 10 V cm⁻¹ for 1 h. There was no ethidium bromide present during the run. Following electro-phoresis the gel was stained in 0.5 µg ml⁻¹ ethidium bromide in TBE running buffer. (b) Lanes 1, 2 and 3 and all conditions are as in (a) except that the gel contained ethidium bromide before loading the samples.*

the mobility of the supercoiled plasmid can be altered substantially depending on whether ethidium bromide is present or not. The molecules in *Figure 7.9a* were run in the absence of ethidium bromide. Notice the position of the lower supercoiled band in relation to the largest λ *Hind*III–*Eco*RI marker. It runs just faster than this 21-kb linear species. When the same samples are run in the presence of ethidium bromide however, the supercoiled band has merged with the nicked circular band, which runs slower than the 21-kb λ *Hind*III–*Eco*RI band. As ethidium bromide dye binds to supercoiled plasmids, it reduces negative supercoiling. As supercoiling reduces, the plasmid becomes less compact and mobility decreases. The ability of ethidium bromide to alter the migration of supercoiled plasmids is sometimes used to confirm the presence of supercoiling in a plasmid preparation.

7.9 Heating λ DNA markers

Restriction digests of bacteriophage DNA are a popular source of electrophoretic markers since it is relatively easy to purify intact the large 48-kb molecule of double-stranded DNA. Moreover its entire nucleotide sequence is known [4]. There is a slight complication however, which derives from the fact that the λ bacteriophage converts from a linear molecule to a circular one by virtue of 12-bp complementary sticky ends called cos sites. These can slowly reanneal, so that following restriction enzyme digestion of λ DNA the two fragments in which the cos ends reside can join together and run as a single band (compare lanes 1 and 2 in *Figure 7.10*). To counteract this, it is frequently stated that samples of λ marker DNA must be heated to melt the noncovalent joining of the two fragments. However the heating must be carefully controlled. Incubating at 65°C for more than a few minutes, especially in low salt buffers that accelerate DNA strand separation (see Section 3.2.1), begins to melt the rest of the molecule with a subsequent loss in resolution and serviceability of the marker preparation. Limit heating to 50°C for 3 min in low ionic strength buffers. There is generally no need to heat DNA samples to melt the overhanging sticky ends following restriction enzyme digestion. Most restriction digests generate sticky ends of much less

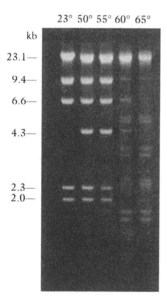

FIGURE 7.10: *Heating λ DNA marker fragments to dissociate the annealing of cos ends. λ bacteriophage DNA markers can display an abnormal pattern when the 4.3-kb fragment joins to the 23.1-kb fragment via the 12-bp single-stranded cos sites. In low ionic strength buffer such as TE (10 mM Tris, 1 mM EDTA) or in this case H₂O, heating for 5 min at 50°C is sufficient to melt this structure and give the correct pattern of λ HindIII fragments. Heating to 60 or 65°C causes extensive denaturation. Reproduced from Gibco-BRL Focus (1994) **16**, 24, with permission from Life Technologies Inc.*

than 12 bp and consequently they are much less likely to reanneal and remain base paired at room temperature. Often restriction enzymes are used or inactivated at elevated temperatures. The presence of higher salt concentrations in the enzyme buffers (50–100 mM) stabilizes the molecules and prevents widespread melting of the double strands. However, take care: unexplained smears or the appearance of new bands in a double-stranded DNA sample that has been heated generally means that the molecules have been partially denatured.

7.10 A note on taking pictures

The most widespread method for recording the results of an ethidium bromide-stained separation is to photograph the gel on a 302-nm transilluminator using a hand-held Polaroid camera (see Section 5.3). There are only two parameters to control: the shutter speed and the aperture setting often referred to as F-stops. To obtain a good photograph the film has to be exposed to an optimum amount of light. Too little and the picture is faint; too much light and the photograph is too bright and it is difficult to resolve details. The amount of light is controlled by the following.

- The length of time the shutter is open: the shutter speed. Longer, slower opening allows more light to strike the film than fast, short opening times. The shutter speeds that are available are usually marked B, 1, 2, 4, 8, 16, 30, 60, 125. These numbers represent the fraction of a second for which the shutter is open. Thus a setting of 8 indicates light falls on the film for $\frac{1}{8}$ of a second. Setting B is a manual control, here the shutter is open for as long as you keep your finger on the trigger (or button). Of course any movement of the camera or the gel during a long exposure will result in a blurred image.
- The size of the hole through which light passes is the aperture. A larger hole lets more light into the camera when the shutter opens. The aperture settings are marked 32, 22, 16, 11, 8, 5.6, 4.5. For the purposes of this explanation it is sufficient to say that the larger the number the smaller the aperture. Thus all else being equal more light will enter the camera at a setting of 4.5 than at a setting of 32. The size of the aperture also determines the depth of focus, that is how much of a three-dimensional image is in focus. High settings (32) (i.e. small apertures) have a greater depth of focus than low settings (4.5) (i.e. large apertures). This is of less concern with a fixed-distance hand-held camera.

A successful photograph requires a compromise between speed and aperture. The majority of the ethidium bromide photographs in this chapter were taken with the camera set on shutter speed B (using an exposure time of 3–4 sec). This requires the camera to be held very steady. The aperture setting was 16. The exact speed and aperture settings will depend on the intensity of light from the transilluminator. The 'glass' filter that overlies the UV tubes gradually 'solarizes' under the influence of UV, becoming irreversibly less transparent in proportion to the number of hours exposure. The visible light photographs were taken on a white light box, with an X-ray film viewer at a shutter speed of $^1/_{16}$ sec and an aperture setting of 32. The film speed or sensitivity also influences the exposure parameters. Polaroid 667 black and white film used for these photographs has an ASA 3000 rating. It is more sensitive, or faster, than the ASA 100 film commonly used for 'holiday snaps'. It is also important to remember that all cameras used to take photographs on a transilluminator should be equipped with a 'deep orange' filter (Wratten 22). The Wratten 22 filter provides optimum contrast for black and white photographs of ethidium bromide gels. A light yellow Wratten 2A filter is often recommended to prevent UV light damaging the Wratten 22. These filters are inexpensive flexible gelatine sheets that can be fixed inside the camera. The Wratten 22 does not interfere too seriously with photographs of gels in visible light, for example Coomassie blue-stained protein gels, which you may want to photograph with the same camera. However, a Wratten 15 deep yellow filter is recommended for maximum black and white contrast of Coomassie blue-stained proteins.

7.11   Recording the position of marker bands when DNA is transferred to a membrane

DNA molecules are separated on gels and transferred to membranes to enable nucleic acid–nucleic acid hybridization to take place between the immobilized target molecule and the probe in solution, for example in a Southern blot procedure (see Section 5.8). Of course size markers such as λ phage restriction digests should be run alongside the DNA sample on the agarose gel to check for proper migration. However, it is vitally important that procedures are adopted that also enable you to record the position of the size markers on the blotted membrane. If this is not carried out, the information that can be obtained from such an experiment can be seriously

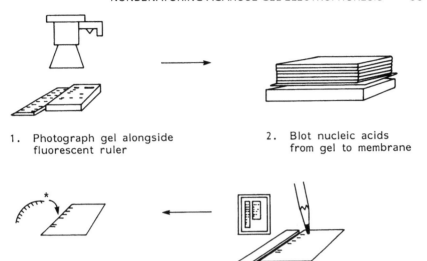

1. Photograph gel alongside
    fluorescent ruler

2. Blot nucleic acids
    from gel to membrane

4. Carry out the detection
    process: for example
    hybridization with a
    labeled probe

3. Indicate the position of the
    markers on the membrane by
    referring to the photograph

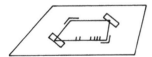

5. If detection requires capturing an
    autoradiographic or chemiluminescent
    image, ensure that the position of
    the membrane on the film is recorded
    by taping and 'drawing' round the
    membrane to scratch the film

FIGURE 7.11: *Recording the position of ethidium bromide-stained
markers on to transfer membranes. It is essential that the positions of
nucleic acid size markers are recorded on to membranes to which
experimental samples have been transferred. Although there are a number
of elegant methods that use labeled markers or probes which will detect
the marker sequences, the use of a fluorescent ruler also accomplishes the
task.*

deficient. For example, it can be difficult to determine whether a band
on an autoradiographic film represents a complete or partial
restriction fragment if you cannot unambiguously determine its size.
The operation is so simple and straightforward that there is no excuse
for not performing it as a matter of routine (see *Figure 7.11*). The
procedure is to photograph the ethidium bromide-stained agarose gel
alongside a UV fluorescent ruler. (It is always a good idea to observe

agarose gels prepared for Southern blots before and after transfer, to check for correct migration and efficient transfer.) UV fluorescent rulers can be obtained from most gel electrophoresis suppliers. The membrane, which is cut to the same size as the gel, can then be marked with a pencil or ball point pen to indicate the positions of the wells and each of the markers by reference to the ethidium bromide photograph. If for some reason you wish to avoid staining the samples in the gel with ethidium bromide and exposing them to UV on the transilluminator for even a brief period, the marker lanes can be sliced off the gel with a scalpel and ruler and stained independently. So long as you keep an eye open for swelling or shrinkage in the two sections, it will be quite simple to mark the membrane unambiguously with the positions of the known-size fragments.

7.12 Research applications: overview

In order to place gel electrophoresis of nucleic acids in the context of its ability to uncover facts about biological systems, I have chosen to devote the remainder of this introductory text to discussions of different electrophoretic techniques as they have been applied and described in published research papers. The papers selected serve as excellent and clear illustrations of the use and interpretation of standard electrophoretic techniques. To demonstrate how electrophoretic techniques are described in scientific journals and to provide an interpretation of the necessarily abbreviated descriptions of gel electrophoresis that appear in published articles, I have also reproduced the original figure legends. The papers are all readily available. Note that a selection of molecular biological terms are given brief explanations in the Glossary.

7.13 Research application. Nondenaturing agarose gel electrophoresis: Southern blotting

7.13.1 Background

As an example of a research application the article by Martiniuk *et al.* (1986). 'Isolation of cDNA for human acid α-glucosidase and detection of genetic heterogeneity for mRNA in three α-glucosidase-deficient patients' has been chosen [7]. This paper describes the isolation of a human cDNA clone which encodes the enzyme acid α-glucosidase (GAA). When this enzyme is missing due to an inherited genetic disease lysosomes fill with 1,4-linked α-glucose polymers that are normally broken down by GAA. Different patients suffer from muscle and liver damage of varying degrees of severity: the disease is heterogeneous. The GAA cDNA was recovered by probing bacteriophage vectors containing cDNA inserts prepared from mRNA expressed in human liver (screening a λgt11 library). The library was screened with an antibody to the purified GAA enzyme. The antibody bound to plaques of the bacteriophage that contained the GAA cDNA, since in *E. coli* the λgt11 vector is able to express cDNAs inserted into the *Eco*RI cloning site. *Figure 7.12* reproduces a figure from the above paper. The image is an autoradiograph of a Southern blot analysis.

7.13.2   Procedure

The following steps were carried out in order to produce the image in *Figure 7.12*.

(1) Total DNA was prepared from mouse cells, human cells or mouse–human hybrid cultured cells which contain specific fragments of human chromosomes.

(2) Each DNA sample was digested with the restriction enzyme *Eco*RI and the resulting fragments of double-stranded DNA were loaded on to a nondenaturing 0.8% agarose gel containing ethidium bromide.

(3) The gel was subjected to an electric field and the fragments were separated according to length. The shortest fragments migrated the furthest in the gel. As so many fragments will be present it is usually impossible to identify individual bands: 'a smear' is how genomic digests are usually described. To identify individual fragments a homologous probe is required.

(4) Following electrophoresis the DNA fragments were blotted out of the gel on to a nucleic acid-binding nitrocellulose filter (a Southern blot). This procedure denatures DNA into single strands so that lengths of single-stranded sequence are available to hybridize with a nucleic acid probe.

(5) The probe, a 2-kb GAA cDNA, was labeled by nick translation with ^{32}P dNTP and hybridized to the Southern filter.

(6) The filter was washed free of unbound probe.

(7) An autoradiogram was prepared by exposing the filter to an X-ray film.

(8) Note that λ *Hind*III digested markers were run on the gel. Their positions were marked on the filter and are indicated in *Figure 7.12*.

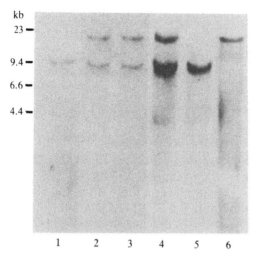

FIGURE 7.12: *Analysis of DNA from human, mouse and human–mouse somatic hybrid cells. DNA was digested with EcoRI, electrophoresed in 0.8% agarose gel, transferred to nitrocellulose filters and probed with the cDNA insert from phage GAA-67. Lane 1: mouse DNA, which showed a 10-kb fragment. Lane 6: human DNA, which showed a 20-kb fragment. Lane 4: DNA from a somatic cell hybrid (6D-19A) containing a pter–q25 segment of human chromosome 17 plus additional chromosomes, showed the 20-kb human fragment. Lane 5: DNA from a 5-bromo-deoxyuridine-resistant, chromosome 17-negative subclone of 6D-19A, which does not show the human fragment. Lane 2: DNA from a somatic cell hybrid (1A-2A) that contains a 17pter–q23 translocation, which also shows the human band. (This hybrid contained only human chromosome 22 and a small pter portion of chromosome 19.) Lane 3: DNA from the most informative hybrid (1A-1A), which contains only a segment (q21–23) of chromosome 17 translocated to a mouse chromosome 17 (and no other human chromosomes), also showed the human 20-kb fragment. At left are positions and sizes of HindIII fragments of bacteriophage λ DNA run in parallel to provide size markers. Reproduced from ref. 7 with permission.*

7.13.3   Interpretation

There is a 10-kb band present in all samples that contain mouse DNA: lanes 1, 2, 3, 4 and 5. Evidently in mice, the human GAA cDNA has sufficient homology to detect the mouse sequence. The human DNA contains a 20-kb *Eco*RI fragment that hybridizes with the GAA probe. The human–mouse cell hybrids in lanes 2, 3 and 4 contains a segment of human chromosome 17, known from previous work to be the location of the GAA gene. Each of these cell lines contains the same 20-kb *Eco*RI fragment that hybridizes with the human GAA cDNA probe. The mouse–human hybrid in lane 5 has been selected in 5-

bromodeoxyuridine. Resistance to this drug requires the loss of the thymidine kinase gene, which is also on human chromosome 17. The loss of the 20-kb human band in this sample is strong evidence that the cDNA does indeed encode human GAA. The presence of the 10-kb mouse band acts as an internal control that good-quality DNA was loaded into this lane.

The importance of nucleic acid electrophoresis to this study was that it permitted the hybridization signal from the mouse 10-kb *Eco*RI fragment to be distinguished from that of the 20-kb human band. This could not for example be accomplished by dot-blot hybridization. The different band intensities are of little consequence. They probably reflect variations in the amounts of DNA loaded on to the gel.

References

1. Sambrook, J., Fritsch, E.F. and Maniatis, T. (1989) *Molecular Cloning: A Laboratory Manual,* 2nd Edn. Cold Spring Harbor Laboratory Press, Cold Spring Harbor, pp. 3–35.
2. Brown, T.A. (1991) *Molecular Biology LabFax.* BIOS Scientific Publishers, Oxford, pp. 255–267.
3. FMC BioProducts Catalog (1995) *Products for Molecular Separations,* Technical Applications. FMC BioProducts, Philadelphia, PA, pp. 68–93.
4. Sanger, F., Coulson, A.R., Hong, G.F., Hill, D.F. and Petersen, G.B. (1982) Nucleotide sequence of bacteriophage lambda DNA. *J. Mol. Biol., 162,* 729–773.
5. Johnson, P.H. and Grossman, L.I. (1977) Electrophoresis of DNA in agarose gels. Optimizing separations of conformational isomers of double- and single-stranded DNAs. *Biochemistry, 16,* 4217–4225.
6. Nathan, M. and Fox, D. (1994) Aberrant migration patterns of DNA molecular size standards on agarose gels. *Focus, Life Technol., 16,* 23–25.
7. Martiniuk, F., Mehler, M., Pellicer, A., Tzall, S., La Badie, G., Hobart, C., Ellenbogen, A. and Hirschhorn, R. (1986) Isolation of a cDNA for human acid α-gluocisdase and detection of genetic heterogeneity for mRNA in three α-glucosidase-deficient patients. *Proc. Natl Acad. Sci. USA, 83,* 9641–9644.

8 Denaturing Agarose Gel Electrophoresis

Applications

Single-stranded DNA molecules: 500–20 000 bases.
* Alkaline agarose gels (see protocols).

Single stranded RNA molecules: 500–20 000 bases.
* Analysis of mRNA molecules.
* Research application: Northern Blotting (see Section 8.1).

Protocols

Molecular Cloning: A Laboratory Manual, Chap. 6, Gel electrophoresis of DNA, Alkaline gels. pp. 6.20–6.21. Chap. 7, Extration, purification and analysis of Messenger RNA from eukaryotic cells. pp. 7.37–7.87 [1].

Molecular Biology LabFax, Chap. 8, Electrophoresis. pp. 260–261 [2].

FMC BioProducts Catalog. Technical Applications. pp. 94–97 [3].

RNA Isolation and Analysis, Chap. 3, Characterisation of RNA size. pp. 47–93 [4].

8.1 Research application. Denaturing gel electrophoresis of single-stranded RNA molecules: Northern blotting

8.1.1 Background

We return to the paper which was used to illustrate the use of non-denaturing agarose gels and Southern Blotting in Chapter 7 (see *Figure 7.12*): Martiniuk *et al.* (1986). 'Isolation of a cDNA for human acid α-glucosidase and detection of genetic heterogeneity for mRNA in three α-glucosidase-deficient patients' [5]. To recap, the article describes the isolation of a cDNA clone encoding the enzyme GAA from a λgt11 library by screening with antibody to purified GAA enzyme. The identity of the clone was confirmed by Southern blotting to genomic DNA obtained from human–mouse hybrid cell lines known to carry human chromosome fragments from the region where the GAA locus had been mapped. The next stage was to examine mRNA from human cells obtained from normal and GAA-deficient patients to measure the length and quantity of GAA mRNA. *Figure 8.1* reproduces another figure from the above paper. The image is an autoradiograph from a Northern blot analysis.

8.1.2 Procedure

The following steps were carried out in order to produce the image in *Figure 8.1*.

(1) Total RNA was prepared from normal or GAA-deficient cultured cell lines. mRNA was isolated by oligo-dT affinity chromatography of total RNA. The vast majority of eukaryotic mRNAs have 3' poly-A tails. These will bind to deoxythymidine oligonucleotides (oligo-dT) coupled to a solid support in the presence of high salt conditions. Pure mRNA (a low percentage of the total RNA in a cell, which is mostly rRNA) can be released by washing in a low salt buffer. (A detailed description of RNA isolation and analysis can be found in refs 1,4.)

(2) Each RNA sample was denatured (unfolded) by the addition of formamide and formaldehyde buffer and the resulting single-stranded species of mRNA were loaded on to a denaturing formaldehyde agarose gel.

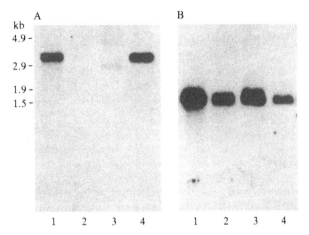

FIGURE 8.1: *(A) Analysis of mRNA from human lymphoblast and fibroblast cells. mRNA (2 µg) was electrophoresed in an agarose/formaldehyde gel, transferred to nitrocellulose, and probed with the cDNA insert from phage GAA-67. Lane 1: RNA from a normal human lymphoid line (GM3201), showing a band at 3.4 kb that hybridizes to the cDNA insert from phage GAA-67. Lane 2: RNA from the GAA-deficient (infantile-onset) fibroblast cell line GM4912, which shows no hybridization to the probe. Lane 3: mRNA from the adult-onset GAA-deficient lymphoid cell line GM1464, which shows a single, less intense band of smaller size. Lane 4: RNA from a second infantile-onset fibroblast cell line (GM244), which shows a band at 3.4 kb. Markers at left show positions and sizes of mammalian rRNA (4.9 and 1.9 kb) and E. coli rRNA (2.9 and 1.5 kb). (B) Duplicate samples were electrophoresed in the same gel, transferred, and probed with a cDNA for adenosine deaminase. All samples exhibited normal-size adenosine deaminase mRNA.*

(3) The gel was subjected to an electric field and the mRNA species were separated according to length. The shortest molecules migrated the furthest in the gel. As so many different mRNAs will be present it is usually impossible to identify individual bands: 'a smear' is how total mRNA preparations are usually described. To identify individual messages a homologous probe is required.

(4) Following electrophoresis the mRNA fragments were blotted out of the gel on to a nucleic acid-binding nitrocellulose filter (a Northern blot). This procedure binds the denatured RNA as single strands so that lengths of single-stranded sequence are available to hybridize with a nucleic acid probe.

(5) The probe, a 2-kb GAA cDNA, was labeled by nick translation with ^{32}P dNTP and hybridized to the Northern filter.

(6) As a control, a duplicate set of lanes were loaded on the same gel, blotted to a different filter and hybridized with a different probe, a cDNA encoding the enzyme adenosine deaminase.

(7) The filters were washed free of unbound probe.

(8) An autoradiogram was prepared by exposing the filter to an X-ray film.

(9) Note that size markers were run on the gel. These were the mammalian 28S and 18S rRNAs and the *E. coli* 23S and 16S rRNAs. Their positions were marked on the filter and are indicated in *Figure 8.1*.

8.1.3 Interpretation

Figure 8.1 is divided into two panels. The left-hand panel (A) shows the hybridization signal when mRNA from a normal human lymphoid line (lane 1) or GAA-deficient fibroblast and lymphoid cultured cell lines (lanes 2, 3 and 4) are probed with the GAA cDNA. Lane 1 shows that the GAA message is approximately 3.4 kb in length. The GAA enzyme was known to be 105 kDa in size (this measurement would have been made using SDS polyacrylamide protein gel electrophoresis). The average size of an amino acid in protein is 110 Da. This suggests that there are in the region of 954 amino acids in the GAA enzyme. As there are three bases per codon, this gives a minimum size for the full-length message of 3 x 954 = 2.86 kb. Since most mRNAs have 5' and 3' untranslated regions it is quite consistent that a 3.4-kb message codes for the GAA enzyme. These calculations are always performed to check whether there is sufficient space within a putative message to code for the protein product it is suspected of encoding. There is clearly only one such message within the cell. Multiple bands would have suggested differential splicing; alternative splicing to create a family of messages. Multiple genes with related sequences would have been detected on the Southern blots. The Northern blot result also shows that the 2-kb GAA cDNA cannot encode the complete gene. Either it contains one of the sites used in its cloning (in this case *Eco*R1 was used to clone the cDNA into the λgt11 bacteriophage) or it became truncated, usually at the 5' end, during the reverse transcriptase construction of the cDNA library. In lane 2 the GAA-deficient sample shows no detectable GAA mRNA, whilst lane 4 from a different GAA-deficient patient shows normal length GAA mRNA. The lane 3 GAA-deficient sample reveals a trace amount of GAA mRNA running at about 2.9 kb.

The discussions of whether or not GAA mRNA was present in one or other of the samples would be meaningless if it were not for the evidence in the right-hand control panel (B). In this case the probe was for adenosine deaminase, which all of the cells should express. The blot shows that there are similar but not identical levels of undegraded mRNA in each sample. Importantly, the small differences in the adenosine deaminase blot do not correlate with those in the GAA-probed Northern blot. In summary, the patients from whose cells the mRNA in lanes 2 and 3 was prepared had GAA enzyme deficiency and very little GAA mRNA. This might indicate a gene deletion, a mutated promoter, splice site or a translational termination codon, all of which could cause a deficiency of GAA mRNA. (Of course large deletions in the gene could be detected using Southern blots.) The patient in lane 4 had normal length GAA mRNA but inactive enzyme. This probably reflects a missense alteration in the gene. Further details of the use of molecular biology techniques in human genetics can be found in ref. 6.

The importance of nucleic acid gel electrophoresis to this study was that it permitted the length of the GAA mRNA to be established. It also tested for the presence of alternatively spliced products and permitted the examination of the size of the mRNA in the GAA-deficient patients. The next stage in such a study would be to try to use the GAA cDNA to probe the cDNA library for longer inserts in the hope that they might contain the full-length cDNA. This would then be used to determine its nucleotide sequence.

References

1. Sambrook, J., Fritsch, E.F. and Maniatis, T. (1989) *Molecular Cloning: A Laboratory Manual,* 2nd Edn. Cold Spring Harbor Laboratory Press, Cold Spring Harbor, pp. 7.37–7.87.
2. Brown, T.A. (1991) *Molecular Biology LabFax.* BIOS Scientific Publishers, Oxford, pp. 260–261.
3. *FMC BioProducts Catalog* (1995) *Products for Molecular Separations,* Technical Applications. FMC BioProducts, Philadelphia, PA, pp. 94–97.
4. Jones, P., Qui, J. and Rickwood, D. (1994) *RNA isolation and analysis.* BIOS Scientific Publishers, Oxford, pp. 47–93.
5. Martiniuk, F., Mehler, M., Pellicer, A., Tzall, S., La Badie, G., Hobart, C., Ellenbogen, A. and Hirschhorn, R. (1986) Isolation of a cDNA for human acid α-glucosidase and detection of genetic heterogeneity for mRNA in three α-glucosidase-deficient patients. *Proc. Natl Acad. Sci. USA,* **83,** 9641–9644.
6. Cooper, D.N. and Krawczak, M. (1993) *Human Gene Mutation.* BIOS Scientific Publishers, Oxford.

9 Pulsed Field Agarose Gel Electrophoresis

Applications

Linear double-stranded DNA molecules: 20 000–2 000 000 bases.

- Genome mapping of restriction fragments and whole chromosomes.

- Research Application: Genome analysis; (see Section 9.4).

Protocols

Molecular Cloning: A Laboratory Manual, Chap. 6, Gel electrophoresis of DNA, Pulsed-field gel electrophoresis, pp. 6.50–6.58 [1].

FMC BioProducts Catalog. Technical Applications, Pulsed Field Gel Electrophoresis Systems, pp. 94–97 [2].

Pulsed field gel electrophoresis: A practical guide. [3].

Methods in Molecular Biology, Vol 31, Physical mapping of the human genome by pused field gel electrophoresis, pp. 121–133 [4].

Methods in Molecular Biology, Vol 31, Field inversion gel electrophoresis, pp. 135–146 [5].

9.1 The principles of pulsed field technology

The fundamental principles of pulsed field technology were outlined in Section 2.3. Essentially, by employing pulses of voltage from different directions and durations, very much larger DNA molecules can be resolved than when using conventional, constant field gels. In constant field gels, all large molecules >30 kb migrate together at an anomalously high rate. In pulsed field gels, these molecules are retarded and can be resolved from each other. This property derives from the requirement that molecules take a period of time to reorient to the new field. The time taken is in proportion to their overall size. Thus when the timing of the pulses is chosen correctly large molecules can be separated according to size.

The development of pulsed field gels has enabled substantial progress to be made in the sphere of genome mapping. This applies equally to the mapping of complex mammalian genomes fragmented through restriction enzyme digestion, and to the characterization of microbial species whose individual chromosomes are often of the order of several megabases (1 Mb = 1 million bases) in size and are amenable to pulsed field resolution in undigested form. The human genome is an estimated 3000 Mb in size.

9.2 Pulsed field electrode geometry

In *Figure 2.9* a simple on–off–on pulse was described. In practice it has been found necessary to use changes in field direction to provide the requirement for a pulsating electric field as this has been shown to give the most effective separations. Considerable research has been carried out to establish the best geometric orientation for the electrodes that supply the pulsating field (see refs 3,6 for a review). The critical test is to be able to achieve straight-line migration without the bands warping excessively in the lower regions of the gel. Deviations from straight-line migration are due to unhomogeneous electric fields within the gel tank. A selection of some of the electrode arrangements in use is illustrated in *Figure 9.1*. Note that at the heart of all pulsed field applications the gel itself is a conventional low-percentage nondenaturing agarose gel formed using conventional buffers such as TBE. Whatever new arrangements of the electrodes

•Transverse alternating field electrophoresis TAFE	•Field inversion gel electrophoresis FIGE

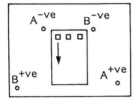

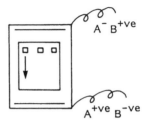

•Requires a vertical gel in a large tank of buffer

•Requires no special gel apparatus. Microprocessor pulse generator can be attached to conventional horizontal gel tanks

•Contour clamp electric field CHEF	•Rotating gel-constant field (once marketed as the 'Hula gel' by Hoefer)

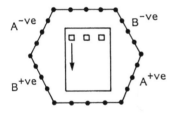

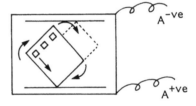

•Requires special gel apparatus but microprocessor controlled point source electrodes give very uniform fields and hence very straight lanes

•The angle and speed of rotation can be altered to effect straight line separation

FIGURE 9.1: *Alternative electrode geometries for pulsed field gel electrophoresis (adapted from ref. 6). Field is switched from A to B and back.*

and pulse regimens are developed in the future, there is one essential fact: the longer the fragment to be resolved, the longer the pulse employed. In practice pulses can vary from 5 to 1000 sec, at field strengths from 2 to 10 V cm^{-1}. Unlike conventional gels, which are relatively immune to changes in running temperature, pulsed field gels need to be run at low temperature (4–15°C) to maintain resolution. Run times range from 10 h to several days.

9.3 Sample preparations for pulsed field gels

Standard methods of DNA preparation can lead to significant shearing or breaking of double-stranded molecules; 200 kb is likely to be the upper limit in most preparations. This is of no consequence when analyzing plasmid and bacteriophage molecules, which will be considerably smaller than 200 kb and, on average, will contain only a low percentage of molecules broken in random positions. Similarly, when restriction digestion is carried out on genomic DNA with enzymes that cut DNA into fragments substantially shorter than 200 kb, such as *Eco*RI or *Bam*HI, a small amount of randomly broken molecules will only contribute to a slight background in subsequent Southern blotting applications. However, when whole chromosomes or very large genomic fragments are to be studied, random breakages that limit the size of fragments to 200 kb represent a serious problem. To overcome this, samples for pulsed field gels are prepared *in situ*, (see *Figure 9.2*). Cells of whatever variety are placed in molten low melting point agarose that is allowed to harden into a small agarose plug. All subsequent lysis and digestion steps are performed within this plug. The agarose protects and retains the long DNA molecules, whilst being permeable to buffers and enzymes into which the plug is placed. Ultimately the plug is located into a well, in the agarose separating gel. From here DNA molecules can move out into the gel under the influence of the applied field, leaving the skeleton of the host cells behind in the plug. The steps involved in preparing high molecular weight DNA for pulsed field electrophoresis are illustrated in *Figure 9.2*. Specially pure grades of low melting point agarose are commercially available for this purpose. The digestion of DNA by restriction enzymes must often be carried out for long periods of time (overnight) with high concentrations of enzymes to allow for the limited rates of diffusion into the agarose plug matrix.

9.4 Research application. Pulsed field agarose gels for genome mapping

9.4.1 Background

As an example of a research application of pulsed field gels I have selected a paper describing the establishment of a physical map of the

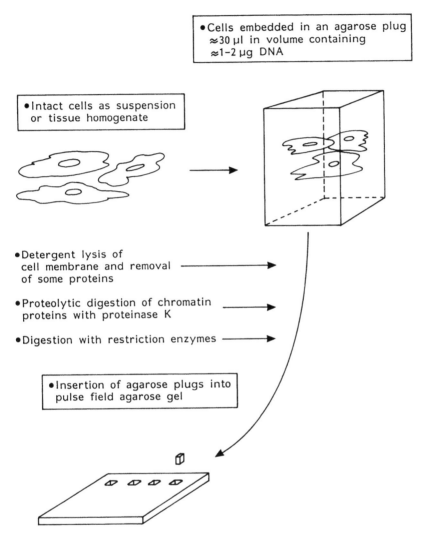

FIGURE 9.2: *The preparation of samples for pulsed field gel electrophoresis.*

genome of the protozoan parasite *Theileria parva* [7]. In Africa, this parasite is carried by a tick and causes the economically important cattle disease East Coast Fever. A genomic map was being constructed to enable sexual recombination between individuals to be followed. Recombination may generate antigenic polymorphism that can assist in the evasion of the host immune system. *Figure 9.3* shows the use of CHEF (contour homogeneous electric field) pulsed field gel electrophoresis to map the parasite DNA. The images are ethidium bromide-stained gels and an autoradiogram of a Southern blot.

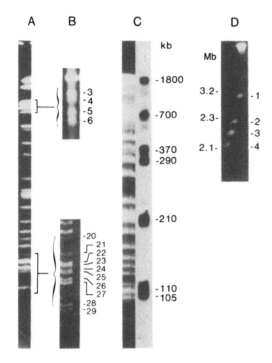

FIGURE 9.3: *Electrophoretic separation of* T. parva *Muguga DNA. (A) Ethidium bromide-stained* Sfi I *fragments separated by CHEF electrophoresis using pulses of 10 s for 16 h followed by 40 s for 3 h, at 10 V cm⁻¹. Fragment numbers are shown on the left. Regions with comigrating fragments (m) and consistently faint bands (f) are indicated. (B) Increased resolution of selected regions using (Upper) pulses of 40 sec for 21 h and (Lower) 3 sec for 21 h. (C) Hybridization of* Sfi I *fragments (Left) with telomeric repeat probe (Right). (D) Separation of chromosomes (numbered 1–4) by using the Pulsaphor apparatus with point electrodes and pulses of 900 sec for 24 h, 600 sec for 24 h, 480 sec for 24 h, and 400 sec for 24 h in 1% agarose/1 × TBE at 3 V cm⁻¹.*

9.4.2 Procedure

The following steps were carried out in order to produce the image in *Figure 9.3*.

(1) A pure suspension of *T. parva* cells was prepared in a low melting point agarose plug and extracted and digested with detergent and proteinase K.

(2) For *Sfi* I restriction enzyme digestion, the agarose plugs were incubated in enzyme, rinsed and positioned in the gel, which consisted of 1% agarose in TBE.

(3) The gels were run at a variety of field strengths and pulse durations depending on the molecules to be separated. In *Figure 9.3A*, the *Sfi* I restriction digest was resolved using a pulse regimen of 10 sec for 16 h and then 40 sec for 3 h both at 10 V cm^{-1}. To expand the resolution of compressed regions where several similar sized fragments migrated closely together, alternative parameters were employed. To resolve a compressed region consisting of fragments in the region of 700 kb a 40-sec pulse was employed for 21 h (*Figure 9.3B*, upper panel). To resolve a compressed region of lower molecular weight, around 100 kb, a 3-sec pulse was employed over 21 h. In *Figure 9.3D*, whole unfragmented chromosomes were analyzed. These gels were run at 3 V cm^{-1} for 24 h with 900-sec pulses, followed by 24 h of 600-sec pulses, 24 h of 480-sec pulses and finally 24 h of 400-sec pulses. A considerable 4-day electrophoretic experiment by anyones' standards!

(4) Following electrophoresis the gels were stained with ethidium bromide and photographed.

(5) In *Figure 9.3C* Southern blotting was carried out using a radioactively labeled oligonucleotide probe with homology to structures at the ends of chromosomes called telomeres. The 28-base synthetic single-stranded DNA molecule was in all likelihood (scientific papers often do not give these minor details) labeled at the 5' end using ^{32}P-γ-ATP and the enzyme polynucleotide kinase. In the case of pulsed field gels, particular care is taken to make sure there is sufficient *in situ* strand breakage by depurination (acid treatment) so that the very large fragments of DNA are broken into small pieces compatible with efficient transfer to the nitrocellulose blotting membrane. Of course this does not compromise the efforts made to produce high molecular weight DNA, because by this stage the molecules have already traveled within the gel according to their size. The fragmentation step in the Southern blotting procedure occurs within the individual bands, which transfer vertically to the nucleic acid-binding filter.

(6) Marker DNAs are crucial for pulsed field gels. The production of a band on a gel under certain conditions is no guarantee that the band is of a particular molecular weight. In this study, markers were prepared by using the enzyme DNA ligase to form multimers of linear λ bacteriophage genomes. This must of course also be carried out in agarose plugs. The result is a ladder of 48.5-kb increments. Another popular source of pulsed field markers are whole chromosomes from *Saccharomyces cerevisiae* or other yeast species. Again these must be prepared by lysing cells in agarose plugs. There are now a number of commercial sources of pulsed field markers. For a laboratory starting out in pulsed field technology, these would appear to be a wise purchase, at least

until you have gained sufficient experience to reproduce the marker preparations with home-made reagents. It is essential that a watchful eye is kept for evidence of band inversions (see Section 2.2.2), because some pulsed field regimes can throw up a lot of these artifacts.

9.4.3 Interpretation

Figure 9.3A (gel photograph) shows that 33 *Sfi* I fragments result from the digestion of *T. parva* DNA. Initially there were two compressed regions where it was suspected that multiple bands were running together, as bands were observed that stained anomalously brighter than bands above and below. By increasing or decreasing the pulse duration, groups of co-migrating bands were resolved (*Figure 9.3B*, upper and lower panels respectively). The 33 *Sfi* I fragments range from about 1800 kb to 100 kb. These fragments would be completely unresolved in a conventional constant field gel.

Having resolved the genome into a number of *Sfi* I fragments, the next question was to ask which fragments represent the ends of the chromosomes. Linear eukaryotic chromosomes have specialized DNA sequences known as telomeres at their ends to ensure proper replication and stability [8]. Telomeres often contain species-specific repeat sequences. An oligonucleotide containing two copies of a 14-base repeat from another protozoan species was synthesized, labeled and used to probe a Southern blot of *Sfi* I digested *T. parva* DNA that had been separated by CHEF pulsed field electrophoresis. The resulting autoradiograph is aligned with the ethidium bromide-stained *Sfi* I digest pattern in *Figure 9.3C*. Eight bands were found to hybridize to the telomeric probe: 1800, 700, 370, 290, 210, 110, 108 and 105 kb. The 110–105 doublet was resolved into a 110–108–105 triplet by running a gel with a shorter pulse frequency. Eight fragments containing telomeres suggests four chromosomes with a telomere at each end. This was confirmed by running undigested DNA on a pulsed field gel. This should contain whole chromosomes. *Figure 9.3D* shows indeed that four distinct bands are seen at 3.2, 2.3, 2.2 and 2.1 Mb after 4 days' electrophoresis. These sizes sum to 10 Mb, which correlates well with the total length of the *Sfi* I digested DNA.

The importance of nucleic acid gel electrophoresis to this study was that it provided the means by which the large 10-Mb *T. parva* genome could be broken down into distinct fragments. Further progress in the assembly of a physical map was achieved by isolating so-called 'linking-clones', fragments of DNA that span *Sfi* I sites, and using these as probes in Southern blotting experiments. Further information was obtained by carrying out *Sfi* I partial digests using a range of restriction enzyme concentrations. The results were analyzed on pulsed field gels and the patterns of appearance and disappearance of fragments allowed the ordering of the *Sfi* I sites in the genome. When polymorphic information can be placed on the genome map, it will enable sexual reproduction and recombination of the parasite to be followed in the host, even in the absence of morphological evidence of these processes.

References

1. Sambrook, J., Fritsch, E.F. and Maniatis, T. (1989) *Molecular Cloning: A Laboratory Manual,* 2nd Edn. Cold Spring Harbor Laboratory Press, Cold Spring Harbor, pp. 6.50–6.58.
2. *FMC BioProducts Catalog* (1995) *Products for Molecular Separations,* Technical Applications. FMC BioProducts, Philadelphia, PA, pp. 94–97.
3. Birren, B. and Lai, E. (1993) *Pulsed field gel electrophoresis: A practical guide.* Academic Press Inc., San Diego.
4. Boultwood, J. (1994) in *Methods in Molecular Biology.* Vol. 31, Protocols for Gene Analysis, (A.J. Harwood, ed.). Humana Press Inc., Totowa, pp. 121–133.
5. Heller, C. (1994) in *Methods in Molecular Biology,* Vol. 31. Protocols for Gene Analysis, (A.J. Harwood, ed.). Humana Press Inc., Totowa, pp. 135–146.
6. Gardiner, K. (1991) Pulsed field gel electrophoresis. *Analyt. Chem.,* **63,** 658–665.
7. Morzaria, S.P. and Young, J.R. (1992) Restriction mapping of the genome of the protozoan parasite *Theileria parva. Proc. Natl Acad. Sci. USA,* **89,** 5241–5245.
8. Blackburn, E.H. (1991) Structure and function of telomeres. *Nature,* **350,** 569–573.

10 Nondenaturing Polyacrylamide Gel Electrophoresis

Applications

Double-stranded molecules 15–300 bases.
- Separation of PCR fragments: Research application: (Section 10.2).
- Purification of oligonucleotides.
- Purification of labeled DNA fragments.
- Bandshift assays. Research application: (Section 10.4).
- Double-stranded conformational polymorphism (DSCP).

Single-stranded molecules 50–300 bases.
- Single-stranded conformational polymorphism (SSCP). Research application: (Section 10.3).

Protocols

Molecular Cloning: A Laboratory Manual, Chap. 6, Polyacrylamide gel electrophoresis, pp. 6.36–6.48 [1].

Molecular Biology LabFax, Chap. 8, Electrophoresis, pp. 261–264 [2].

Methods in Molecular Biology, Vol. 31, The gel retardation assay, pp. 339–347 [3].

Appendix to AT Biochem Handbook (1994) Protocols and Bibliography for SSCPs and DCSPs [4].

Methods in Molecular Biology, Vol. 31, A nonradioactive method for the detection of single-strand conformational polymorphisms, pp. 205–210 [5].

10.1 Overview

The greater complexity of preparing and running polyacrylamide gels when compared with agarose gels means there is a tendency in many laboratories to restrict the use of polyacrylamide to large-dimension denaturing gels such as those used for DNA sequencing (see Chapter 11). Agarose gels, whether they are denaturing or nondenaturing gels, can be used for most electrophoretic procedures, but polyacrylamide is capable of superior performance in the low molecular weight ranges. The preparation of polyacrylamide and agarose gels is reviewed in Chapter 4.

Consider the resolution of double-stranded DNA fragments of less than 500 bp in length. Agarose gels have to be prepared at quite high concentrations, in the region of 2%, to resolve very short fragments successfully. These gels can take many hours to run. They can also be quite opalescent, which reduces the sensitivity of ethidium bromide detection. Bands can become broad and ill resolved. Moreover, high concentrations of agarose can interfere with DNA extraction procedures (see Section 4.3). In contrast, nondenaturing polyacrylamide gels give very good resolution of low molecular weight fragments, the DNA is easy to extract and is of high purity. Once the gels have been removed from between the glass plates used to prepare vertical poylacrylamide gels, ethidium bromide staining can be used if there is a reasonable level of nucleic acid in the bands. If levels of DNA are very low, then silver staining can be used to provide a highly sensitive detection system. Failing this, radiolabeling works extremely efficiently. Gels can be exposed to film when wet if ^{32}P is used. For ^{35}S or ^{33}P, polyacrylamide gels must be dried to eliminate quenching of these isotopes.

In response to the drawbacks associated with the electrophoresis of DNA fragments of 500 bp and less in agarose gels, there is a lot of activity in commercial companies aimed at providing an electrophoresis matrix that works like polyacrylamide but pours with the ease of agarose (see Section 4.4). Notwithstanding these efforts, this section is a reminder not to forget the very high resolution for small double-stranded fragments with even modest sized (10 cm x 8 cm) nondenaturing polyacrylamide gels. This application is particularly important in the analysis of small-sized PCR fragments. The verification of the number of amplified products and their exact size is often crucial to accurate assessment of the experiment. A research application involving nondenaturing polyacrylamide gels to analyze double-stranded PCR fragments is described in the next section. Later sections describe the use of nondenaturing polyacrylamide to examine mutations and protein-binding sites.

10.2 Research application. High-resolution separation of double-stranded DNA fragments

10.2.1 Background

A nice example of the power and utility of the PCR process is provided in a paper from Li, Cui and Arnheim (1990) [6] to determine which of a number of polymorphic alleles (versions of a gene) are present in a single human sperm. The technique relies on the high-resolution separation of small double-stranded DNA PCR fragments. These are produced in an ingenious manner so that small differences in their length indicate which of two alleles are present.

The advantage of being able to detect which allele is present in a single sperm is that it facilitates mapping genes by determining the recombination frequencies between loci marked by alternative alleles. Hitherto, this is something that can only be achieved by several generations of breeding: a time-consuming experiment in long-lived animal species. Moreover, in humans, mapping loci is severely hampered by the tendency to have relatively few offspring and by the inability to direct the crosses! In this method, starting with an individual who is heterozygous at a number of linked loci, recombination frequencies can be determined from a single sample of semen. If individual sperm can be analyzed reliably, each sperm is equivalent to an individual offspring in the data it provides. Geneticists will recognize that several linked polymorphic genes could be ordered on the chromosome by the low frequency of the multiple crossover class.

The approach taken in the study was to analyze three loci for which well-characterized, single base change polymorphic alleles were known. These were RFLPs in the parathyroid hormone (PTH) gene, in the Gγ globin gene (Gγ, HBG2), both on chromosome 11, and the low-density lipoprotein receptor (LDLr) on chromosome 19. To determine which allele was present, a fragment spanning the polymorphic fragment was amplified using a pair of PCR primers. A sample of this reaction was then subjected to a second amplification using one primer from the first reaction and a third primer located inside the original return primer, and whose 3' terminus was complementary to either one or other of the polymorphic alleles. By altering conditions so that a primer with a mismatched 3' terminus was not amplified but the matched primers amplified successfully, the identity of the allele that had amplified could be identified by the length of the final product. This was because primers with alternative 3' termini had different 5' ends: the primers were a total of either 20 or 35 bases in length. The gel electrophoretic procedure had therefore to be able to distinguish PCR products that differed by just 15 bases. This method is called allele discrimination by primer length (ADPL).

The technique of reamplifying a PCR reaction with a second pair of primers located inside the first pair is called *nesting*. The variation in this paper, where only one of the primer pair is replaced, is called *hemi-nesting*. Nesting is a very useful method to convert a messy, nonspecific PCR into a clean one. Alternatively, it can be used to confirm that the band observed on a gel really is the fragment you think it is. The principles of the ADPL method are illustrated in *Figure 10.1*.

10.2.2 Procedure

(1) A single sperm was added to a PCR reaction containing three pairs of primers: one for PTH, one for Gγ and one for the LDLr locus. The single sperm was isolated using a micromanipulated pipette observed under a phase contrast microscope.

(2) Following PCR amplification, three samples of the first reaction were individually amplified using a hemi-nesting PCR where the nested primer is a mixture of two primers differing by a single base at the 3' end that corresponds to the known site of polymorphic variation. These primers also differ in length by 15 bp so that they can be distinguished in the ADPL procedure (see *Figure 10.1*).

(3) After the second amplification, the products were applied to a 10 cm x 8 cm x 0.15 cm nondenaturing polyacrylamide gel. The total polyacrylamide concentration was 8%T; the ratio of acrylamide monomer to bis-acrylamide cross-linker was 19:1. This ratio gives the smallest pore size (see Section 4.5) and is the most commonly used matrix for nucleic acid polyacrylamide gels, whether they are used for denaturing or nondenaturing applications. Electrophoresis was carried out at 100 V for 1.5 h.

(4) Subsequent to electrophoresis, the gels were stained in ethidium bromide and photographed on a transilluminator. Size markers were provided by a *Msp* I restriction enzyme digest of plasmid pBR322.

The ADPL Procedure

1 PCR using locus specific primers

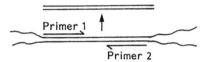

2 Hemi-nested PCR using primers specific
for a single base polymorphism

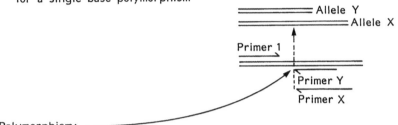

Polymorphism:

3' **A** G T A......   Primer for allele X   35 bases
.......A G **T** C A T......   Template allele X

3' **C** G T A......   Primer for allele Y   20 bases
.......A G **G** C A T......   Template allele Y

Alleles are distinguished by nondenaturing polyacrylamide gel
electrophoresis

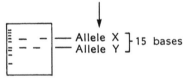

FIGURE 10.1: *The principles of allele discrimination by primer length
(ADPL): high-resolution separation of double-stranded DNA fragments on
nondenaturing polyacrylamide gels.*

10.2.3   Interpretation

There are two images in *Figure 10.2* that depict the simultaneous
analysis of three polymorphic loci in a single human sperm. The upper
image is the ethidium bromide-stained nondenaturing polyacrylamide gel. The lower image is a dot-blot probed with allele-
specific oligonucleotides (ASOs) to verify the performance of the
ADLP procedure. The ethidium bromide-stained gel shows that in
each of lanes 2–8 there are only three bands. In each case one or other

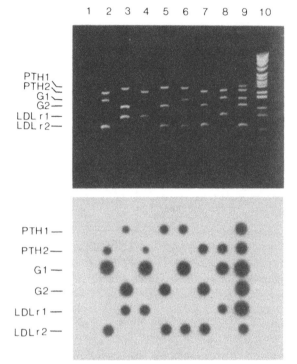

FIGURE 10.2: *Determination of the allelic status at three loci in a single sperm by using the ADPL procedure. Upper panel: ethidium bromide staining. Lower panel: parallel experiment using the ASO hybridization approach to confirm the ADPL allele-typing results. Lane 1: negative PCR control, which received all reagents except a sperm. PTH1 and 2 are the alleles at the PTH locus, lacking and containing the* Taq I *recognition sequence, respectively (termed* Taq I⁻ *and* Taq I⁺, *respectively). G1 and 2 are the* HindIII⁺ *and* HindIII⁻ *alleles at the Gγ locus. LDLr1 and 2 are the* Ava II⁻ *and* Ava II⁺ *alleles at the LDLr locus. Lanes 2–8: single sperm samples. Lane 9: ADPL products from a 3-μl sample of semen from the triply heterozygous sperm donor. Lane 10: pBR322 digested with* Msp I. *Reproduced from ref. 6 with permission from the authors.*

of the PTH, Gγ or LDLr alleles is present, indicating that only a single haploid sperm was introduced into each reaction. The size of each band is indicated to the left of the gel: 172 vs. 157, 139 vs. 124 and 106 vs. 91 bp. The 15-bp additions to the primers with 3' mismatches specific for the polymorphic RFLP sites resulted in clearly distinguishable double-stranded DNA fragments on the non-denaturing polyacrylamide gel. The sample in lane 9 contained a mixture of sperm from the donor who was heterozygous (had both alleles) at each of the three loci. In this sample, all six PCR fragments were present. Examination of the gel reveals that individual sperm showed predominantly either the PTH2/G1 or the PTH1/G2 combinations, but that they can have either LDLr1 or LDLr2 alleles. This is entirely as expected, since PTH and Gγ are linked (are on the same chromosome) whereas LDLr is on a different chromosome. In lanes 6 and 7, recombinant sperm with the PTH1/G1 and PTH2/G2 arrangement were found, indicating a crossover during meiosis. The lower panel in *Figure 10.2* shows the results of the dot-blot study in which PCR amplifications of individual sperm were probed with radiolabeled ASOs. The results of the two methods are entirely consistent. However, the electrophoretic method is considerably less demanding.

Gel electrophoresis was essential for this study to provide, in combination with an elegant PCR procedure, an unambiguous discrimination between two DNA targets that differ by a single base. Small-scale (10 cm x 8 cm) nondenaturing polyacrylamide gels gave the high resolution required. Such a separation would have been difficult to produce with agarose gel electrophoresis.

10.3 Single- and double-strand conformational polymorphisms: SSCP and DSCP

10.3.1 The principle of using gel electrophoretic conformational polymorphisms to detect mutations

A large fraction of the current effort in the medical sciences to acquire and apply the techniques of molecular biology is devoted to the screening of individuals for evidence of mutations in previously characterized genes. Mutations are changes in DNA sequences that may alter the expression of genetic information. Mutations could be deletions, insertions or rearrangements involving many bases, or changes in the identity of single nucleotides. Mutations may be in the germline, for example in inherited genetic diseases where the alteration is present in every cell in the body. Alternatively, mutations can be in somatic cells, for example in mutations leading to cancer, where perhaps only a single clone of cells arises with the DNA alteration.

Deletions or insertions involving large numbers of bases can be immediately recognized using Southern blotting or PCR techniques. On the other hand, where point mutations are the origin of genetic disease, searching for single nucleotide changes in a gene sequence that may contain several thousand bases is literally like looking for a needle in, if not a haystack, then certainly a modest pile of straw. Of course, DNA sequencing is a guaranteed method for finding mutations. However, DNA sequencing, despite the advances in automated gel reading (see Chapter 5) is too substantial a procedure to warrant sequencing very many different individuals. Often there are several candidate genes in which a mutation could be present. Thus many genes would be sequenced without finding a nucleotide change. What is required is a method that will flag a particular individual as having, or not having, a mutation somewhere in a particular gene. This limits significantly the number of genes that have to be sequenced. Two of the most widespread techniques rely on the altered mobility in nondenaturing polyacrylamide gel

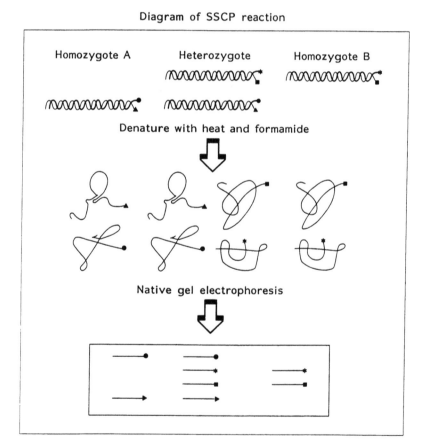

FIGURE 10.3: *The principles of single-strand conformational polymorphism (SSCP) analysis. Reproduced from* AT Biochem Catalog & Handbook *(1994) published with permission from FMC BioProducts.*

electrophoresis that results from different conformations as either single strands or double strands of DNA. These are known as SSCPs and DSCPs.

SSCP. The principle of SSCP is illustrated in *Figure 10.3*. When DNA fragments from a homozygote (an individual where both chromosomes contain the same sequences within the region in question) are denatured by heating and exposure to the denaturant formamide, the two strands are separated (see Chapter 3). If this sample is now cooled rapidly and applied directly to a nondenaturing polyacrylamide gel, the single strands will fold into a stable conformation using intramolecular base pairing. As the two strands of a double-stranded molecule will have very different primary sequences, unless they are an exact inverted repeat, the two molecules may well have different three-dimensional conformations. These different conformations will result in different mobilities in a nondenaturing polyacrylamide gel and thus two bands are generally seen when a double-stranded molecule is subjected to the SSCP process. (If the gel was a denaturing gel, containing high concentrations of urea, the single strands would not be permitted to undergo folding and both strands would migrate together, strictly according to the number of bases in their sequence, but with no regard to the nature of the sequences themselves. Applications employing denaturing polyacrylamide gels, such as DNA sequencing, are described in Chapter 11.)

In *Figure 10.3*, two different homozygote DNA double strands are illustrated: A and B. Although they have the same molecular length, differences in their sequences mean that when subjected to the SSCP process each of the two pairs of strands migrates with a unique mobility. Thus the presence of a mutation, or mutations, that cause A and B to differ can be identified by the observation of altered mobilities of single strands on a nondenaturing gel. In a heterozygous individual, where one chromosome contains A and the other chromosome harbors the B sequence, four different bands are observed on the nondenaturing gel, alerting the presence of a mutation somewhere in the fragment. Of course it is not possible to know where, or what, the mutation is without determining the DNA sequence of the mutant strand. It should also be emphasized that there is no guarantee that a particular mutation will cause an altered migration pattern. Of course it is always vital to run known wild-type DNA on the same gel to be sure of making a valid assessment. A second important control is to run wild-type DNA that has not been subjected to any denaturation. This marks the mobility of the double-stranded form and identifies any single strands that manage to reanneal during the process of gel electrophoresis.

Single-stranded DNA is bound poorly by ethidium bromide (see Chapter 4). For this reason SSCP analyses need to employ a different detection system. The first reports of SSCP used restriction enzyme digests of genomic DNA [7]. These were denatured, run on nondenaturing polyacrylamide gels, blotted on to filters and probed with radiolabeled homologous sequences. Most investigators now use PCR to amplify short, 100–500 bp, regions of genomic DNA or cDNA prepared from reverse transcription of mRNA (RT–PCR). Often primers are positioned to amplify individual exons, which can then be screened for mutations by SSCP. PCR fragments can easily be labeled by including a radiolabeled nucleotide in the amplification reaction. Short fragments are used in SSCP analyses because the contribution that a point mutation can make to the conformation of a single strand is greater the shorter the strand. However, shorter fragments imply greater labor in screening an entire gene for the presence of point mutations. Fragments in the range 100–500 bp are therefore a compromise. Although the resolution of conformational variants that indicate point mutations is unpredictable, the nature of the polyacrylamide matrix can affect the chances of detecting SSCPs. Alternative gel matrices such MDE (AT Biochem, see Chapter 4) have already been developed. Small quantities of mild denaturants can also improve the chances of uncovering bands with altered migration. SSCP is conducted on large-format gels 30–40 cm long. Run times in excess of 10 h are not uncommon.

DSCP. The principle of DSCP, sometimes called heteroduplex analysis, is illustrated in *Figure 10.4*. When DNA from a homozygous individual is denatured with heat and then allowed to reanneal by slow cooling, naturally a single species of double-stranded DNA is reformed. In contrast, when DNA from a heterozygous individual, where one chromosome contains a gene with a slightly different sequence, is allowed to reanneal, the four individual strands can associate at random. Where reannealing between a mutant and a wild-type strand takes place, a mismatch will be present in the double-stranded heteroduplex. In nondenaturing conditions this can give rise to altered migration in polyacrylamide gels. Research suggests that the DNA bends, or kinks, at the site of the mismatch [8]. Although there are four combinations of single strands in DSCP analysis of a heterozygous individual (wild-type/wild-type, mutant/mutant, wild-type/mutant, mutant/wild-type), often the two nonmismatched hybrids will run as a single band. The two mismatched heteroduplexes may, or may not, run as independent bands. Thus the detection of a band with altered mobility is evidence that a mutation is present, but the failure to detect a DSCP or SSCP is not evidence that no mutation is present.

Diagram of heteroduplex reaction

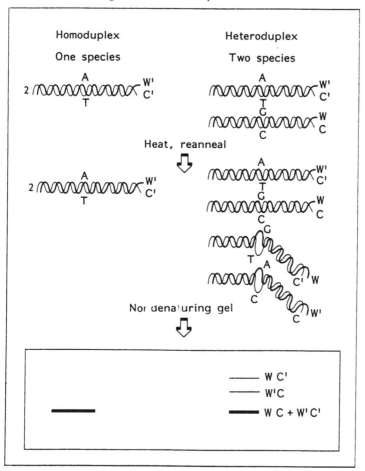

FIGURE 10.4: *The principles of double-strand conformational polymorphism (DSCP) analysis. Reproduced from* AT Biochem Catalog & Handbook *(1994) published with permission from FMC BioProducts.*

Unlike SSCP, ethidium bromide staining can be employed to detect DSCPs, since ethidium bromide has a much higher affinity for double-stranded than single-stranded DNA. Although polyacrylamide does quench ethidium bromide fluorescence to some extent, when PCR is used to generate the double-stranded DNA fragments for DSCP analysis usually it is not difficult to ensure that a sufficient level of DNA is loaded on to the gel to compensate for this. Alternatively, silver staining can be used to obtain very sensitive detection, although this takes a little longer to perform and the DNA fragments cannot be recovered from the gel for further analysis, for example direct PCR-based DNA sequencing [5]. In common with SSCP analysis, short fragments (200–600 bp) offer the most efficient means of detecting mutations by DSCP. Large-dimension, vertical, nondenaturing polyacrylamide gels are used for DSCP analysis. These are frequently run for many hours to achieve the appropriate separation. Gel temperature, ionic strength and the inclusion of small quantities of denaturants have all been shown to affect the likelihood of detecting bands with altered mobilities [8].

10.3.2 Research application. Nondenaturing polyacrylamide gel electrophoresis: using SSCP analysis to detect mutations

Background. The paper chosen to illustrate the application of SSCP analysis, Dumaz *et al.* (1993) 'Specific UV-induced mutation spectrum in the p53 gene of skin tumors from DNA-repair-deficient xeroderma pigmentosum patients' [9]. This study is typical of a great many studies carried out in recent years that have examined the mutation spectrum in somatic cells (cancers) and the germline (inherited genetic diseases). The basis for tumor development is known to be a multistage process (see [10] for a review). One stage encompasses the activation, usually by mutation, of one or more cellular oncogenes. Another step is the loss, or altered expression, of one or more tumor suppressor genes. This paper examines samples of skin tumors for mutations in the tumor suppressor gene *p53*. The tumors were obtained from individuals suffering from the inherited disease xeroderma pigmentosum (XP). This disease is the result of inheriting two defective copies of a gene, specifying a protein that participates in the repair of DNA which has been damaged by the UV component of sunlight.

In order to avoid having to clone and sequence all of the *p53* genes from the 43 tumors collected in this study, genomic DNA from 36 of the tumors was subjected to PCR and analyzed for SSCPs. In the case of mutations in *p53*, this approach represents a considerable advantage because it is known that single base changes can lead to an altered p53 protein, which acts in a co-dominant manner and is permissive for tumor development. The normal allele is therefore present in most tumors and this has to be distinguished from potential mutant alleles (versions) of *p53*. The same reasoning applies to screening for 'carriers' of recessive mutations determining inherited conditions. *Figure 10.5* shows examples of SSCP gels where altered mobilities were detected in DNA fragments amplified from the *p53* gene.

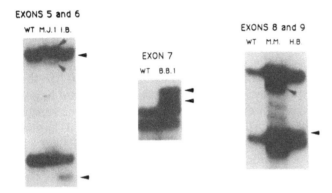

FIGURE 10.5: *SSCP analysis of some amplified exons from various XP skin tumors. The presence of shifted bands (arrowheads) compared to wild-type (WT) sample indicates the presence of at least one mutation on the DNA sequence; initials correspond to patients. Reproduced from ref. 9 with permission from the National Academy of Sciences of the USA.*

Procedure

(1) Genomic DNA was purified from tumors and normal skin from both XP patients and non-XP individuals. PCR was performed using primers that amplified exons 5–6, exon 7 and exons 8–9. α-^{32}P dCTP was included in the PCR amplification to label the DNA uniformly.

(2) Following PCR the samples were denatured with formamide, heated to 100°C for 5 min, chilled on ice and loaded on to MDE vinyl polymer gels (AT Biochem, which was acquired by FMC BioProducts in 1995, see Section 4.5). Electrophoresis was conducted at 8 W constant power at room temperature for between 12 and 24 h.

(3) Autoradiography was carried out and SSCP bands were identified, cut from the gel, eluted and sequenced using direct-PCR methodology. In this way the mutant alleles can be identified unambiguously. Note that this would not be as straight forward with DSCP methods since here the altered bands that indicate the presence of a mutation are a heteroduplex containing both wild-type and mutant strands. Direct sequencing would yield a mixed product.

Interpretation. *Figure 10.5* shows three autoradiographic images of nondenaturing polyacrylamide gels. Each reveals examples of SSCPs from XP tumor DNA in exons 5–6, exon 7 and exons 8–9. To the left of each gel, the pattern from the wild-type *p53* gene is shown. This negative control is absolutely essential for every gel run to screen for

SSCPs. In each case two bands are present, but the separation between the two bands is different in each of the exons. This is because the DNA fragments will be of different length and composition. They would therefore be expected to show different patterns of gel mobility when present as single strands. In the subsequent lanes identified by XP patient initials, a number of bands with altered mobility are found. These exhibit either faster or slower migration. Arrows indicate the presence of SSCP 'shifted bands' (these are not the same as bandshifts described in Section 10.4, which indicate the presence of protein binding to DNA fragments). In some cases, more than one extra band is found. This may indicate one of the following: (i) that both chromosomes carry independent mutations; (ii) that the sample is a mixture of cells which contain either, but not both, mutations; or (iii) that there are two stable conformations attainable by one strand containing a single mutation. These three possibilities can only be resolved by sequencing DNA eluted from the different bands

Gel electrophoresis was essential for this study. The ability to detect the presence of a mutation in a defined region of a gene by altered mobility of a single strand in a nondenaturing polyacrylamide gel saved a considerable amount of unproductive sequencing of nonmutated exons and wild-type alleles. Some 40% of the XP tumors contained mutations in the *p53* gene. When the relevant regions were sequenced, 61% of these mutations were of the CC→TT type, a signature of UV-induced mutations that XP cells cannot repair.

10.4 Bandshift or gel retardation assays

10.4.1 The principles of the bandshift assay

Generally the presence of significant amounts of protein in DNA or RNA samples is not conducive for electrophoresis with good resolution. Proteins that bind to DNA or RNA prevent them from entering the gel matrix or restrict their ability to migrate freely and induce smearing. This is in addition to the fact that if proteins are present they may contain unwanted nuclease activities. In the light of these two considerations, nucleic acids are usually relatively free of protein by the time they are subjected to electrophoresis. However, if comparatively short fragments of nucleic acid are used, specific retardation of gel mobility can be used to gather information on the DNA- or RNA-binding proteins and the sequences with which they interact.

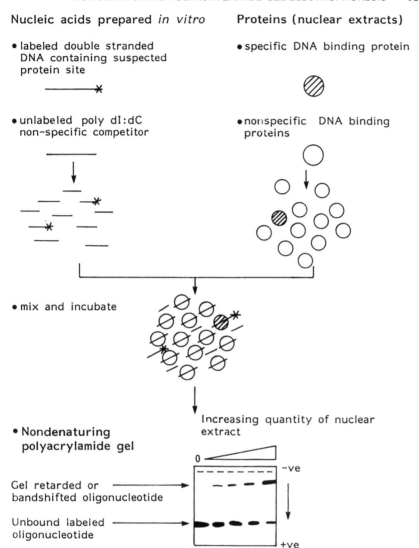

Nucleic acids prepared *in vitro*

- labeled double stranded
  DNA containing suspected
  protein site

Proteins (nuclear extracts)

- specific DNA binding protein

- unlabeled poly dI:dC
  non-specific competitor

- nonspecific DNA binding
  proteins

- mix and incubate

Increasing quantity of nuclear
extract

- **Nondenaturing**
  polyacrylamide gel

Gel retarded or
bandshifted oligonucleotide

Unbound labeled
oligonucleotide

FIGURE 10.6: *The principle of bandshift or gel retardation assays.*

The principles of this gel retardation or bandshift assay are
illustrated in *Figure 10.6*. A short, labeled, double-stranded DNA
fragment, for example a pair of complementary oligonucleotides that
have been allowed to anneal together, will migrate rapidly through a
nondenaturing polyacrylamide gel as a discrete band. If the sequence
contains the recognition sequence for a DNA-binding protein, a
transcription factor perhaps, the binding affinity will generally be
strong enough to resist the force of the electrical field. The increased
cross-sectional area and the reduction in flexibility of the nucleic acid

will lead to a retardation in the mobility of the DNA. The 'band' detected in the absence of protein is 'shifted' further up the gel. The bandshift assay refers therefore to studies of nucleic acid–protein interactions using the technique of altered gel mobilities. Other names for these assays in common use are EMSA (electrophoretic mobility shift assay) or GEMSA (gel electrophoretic mobility shift assay). If more than one protein binds to a particular sequence or a second protein binds to a protein already in contact with a recognition sequence, a further reduction in mobility can occur. These incremental retardations are known as 'supershifts'.

Naturally, the ability of a protein to recognize a specific sequence within a section of DNA requires that both protein and nucleic acid are in nondenaturing conditions. Generally, the DNA must be double stranded and free to partake in intramolecular base pairing. Proteins must be in physiological ranges of ionic strength and pH. Mg^{2+} or Ca^{2+} ions may be required. Oxidation or reduction of thiol groups could be important. Metal ions such as Zn^{2+} can be critical: the integrity of the so-called 'zinc-finger' DNA-binding protein motif is dependent on Zn^{2+} [11]. Such components may need to be present in the gel and running buffer to maintain DNA–protein interactions.

Specific bandshifts can be masked by a mass of nonspecific DNA–protein associations that lead to nondescript smearing. These unwanted interactions can often be minimized by adding an excess of an unlabeled nonspecific competitor (see *Figure 10.6*). Single strands of poly dI:dC (deoxyinosine:deoxycytosine) will anneal to themselves and form a double-stranded competitor. Inosine is a purine base with broad recognition properties. Confirmation that a bandshift is in reality the consequence of a specific interaction should always be obtained by performing unlabeled or cold-competitor studies, the 'cold' meaning unlabeled as opposed to 'hot', which is often used to indicate radioactive labeling. If the bandshift observed is real, then pre-incubating in the presence of an excess of unlabeled DNA of the same sequence should prevent formation of the bandshift. The amount of unlabeled competitor that needs to be added will depend on the quantity of protein present in the incubation. The bandshift, the protein–DNA interaction, still occurs but is now conducted by unlabeled components that are invisible on the autoradiogram of the gel. If this treatment does not eliminate the appearance of the band with altered mobility, then the veracity of the proposed interaction is immediately called into question. *Figure 10.7* shows an example of a bandshift assay from a published research paper, which used unlabeled competitor to validate a bandshift analysis.

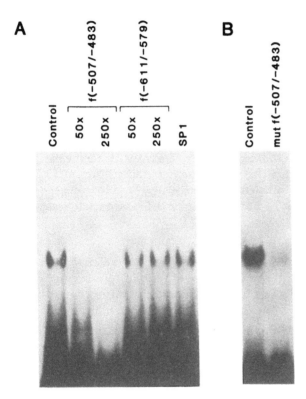

FIGURE 10.7: *Gel retardation analysis of nucleoprotein complexes found with f(–507/–483). (**A**) f(–507/–483) (CCGGGAAGTGGTGGGGGGAGGGAGC) was end-labeled and incubated with 10 µg of MCF-7 nuclear proteins. Unlabeled f(–507/–483) and f(–611/–579) were added at a 50- or 250-fold molar excess compared to the labeled probe. The unlabeled oligonucleotide corresponding to the consensus sequence for SP1 was added at a 50-fold molar excess. (**B**) f(–507/–483) and a mutated fragment at positions –496 and –495 (CCGGGAAGTGG<u>C</u>AGGGGGAGGGAGC) were end-labeled (both 1×10^5 cpm, 5×10^4 cpm/ng) and incubated with MCF-7 nuclear proteins. Reproduced from ref. 12 with permission from the National Academy of Sciences of the USA.*

10.4.2 Research application. Bandshifting assays

Background. As an example of the application of bandshifting, I have selected a paper that characterizes the transcriptional regulation of a gene expressed at high levels in human breast cancer cells. Abe and Kufe (1993) 'Characterisation of cis acting elements regulating the transcription of the human DF3 breast carcinoma associated antigen (*MUC1*) gene' [12]. The DF3 protein is a high molecular weight glycoprotein encoded by the *MUCI* gene. The complete genomic sequence had been determined for more than 1600 bases upstream from the transcription start site. The study describes the construction of deletion variants of this upstream region fused to the reporter gene chloramphenicol acetyltransferase (CAT) in order to map the important areas of the promoter region. Having focused on a length of sequence that appeared important for transcriptional activity in MCF-7 breast carcinoma cells, attention was turned to discover where exactly the regulatory proteins might bind within this region. This was achieved using DNase I footprinting and gel retardation assays (DNase I footprinting assays will be covered in Chapter 11 using data from this same research paper). *Figure 10.7* is an autoradiograph of a nondenaturing polyacrylamide gel used to map protein-binding sites by bandshifting or gel retardation assays.

Procedure

(1) A double-stranded fragment containing a region suspected of containing a regulatory site was prepared by mixing two complementary synthetic oligonucleotides, heating and then cooling slowly to allow the strands to anneal. The molecules were designed with overhanging 5' ends that were filled in by DNA polymerase I in the presence of α-^{32}P dNTPs.

(2) Nuclear extracts were prepared by gently lysing cells, centrifuging out the nuclei and lysing these to release nuclear proteins. Protease inhibitors are generally present throughout these procedures.

(3) Nuclear protein (10 µg) from the breast cancer cell line expressing DF3 was incubated with 10 µg of poly dI:dC for 30 min; 1 ng of ^{32}P-labeled double-stranded oligonucleotide was added and incubated for a further 30 min in a buffer conducive to protein function.

(4) The material was loaded on to a nondenaturing polyacrylamide gel. Bandshift assays can be run on small-sized 10 cm x 8 cm vertical gels. Such gels are run with buffers and conditions typical for nucleic acid electrophoresis rather than those typical for protein gels. Following electrophoresis, autoradiograms were prepared from the dried gel.

Interpretation. In *Figure 10.7A* the position of a bandshifted oligonucleotide complex is seen approximately halfway up the gel. The large quantity of label running lower down the gel represents an excess of oligonucleotides that have not bound nuclear proteins, or which have insufficiently strong interactions to resolve into the discrete band. Poly dI:dC was present in all samples to restrict nonspecific binding. The shifted band indicates that some protein or proteins in MCF nuclear extracts can bind to sequences that lie in the region 507 to 483 bases upstream from the transcriptional start site.

(The fragment is described as f(–507/–483)). By convention the start of transcription is designated +1. Sequences upstream of this, the promoter region, are given negative numbers counting from +1. Positive numbers are within the sequence specifying the transcript.

To prove that the bandshift is a genuine phenomenon, lanes 2 and 3 indicate the effects of 50-fold and 250-fold molar excess of the same unlabeled double-stranded oligonucleotide. The bandshift disappears. Since the prediction is that the signal will be affected, this represents a positive control. Note how the high concentration of unlabeled competitor reduces the smear that extends up the gel to the genuine shifted band. This suggests that the smear also represents sequence-specific binding, but in a fashion where multiple or unstable conformations prevent the resolution into a homogeneous band. Lanes 4 and 5 indicate the effect of competition with unlabeled double-stranded oligonucleotides corresponding to a section of upstream sequences between –611 and –579 of the gene for DF3. There is little diminution of the bandshift signal. This suggests that there is no comparable protein-binding site on this fragment. This represents a negative control. The lane marked SP1 demonstrates that the bandshift is not caused by the well-characterized transcription factor SP1, since competition with unlabeled oligonucleotides containing sites for SP1 do not compete with the bandshift. (There are sequences that resemble the SP1-binding site between –507 and –483.) A further experiment was conducted to explore the DNA-sequence specificity of the bandshift. The result is shown in *Figure 10.7B*. The standard bandshift is shown again in the left-hand lane. The next lane demonstrates the relatively poor binding obtained when the standard labeled oligonucleotide is replaced by one that has two nucleotide changes out of 25 (see figure legend for details). In this way a clear picture of the interaction between DNA sequences suspected of being binding sites for regulatory proteins can be examined step by step, using a relatively simple procedure. Further interactions between DNA sequences and proteins can be explored using a technique called DNA footprinting. This technique represents an application of denaturing polyacrylamide gel electrophoresis and an example from the same study is considered in Chapter 11.

Nucleic acid gel electrophoresis is put to powerful yet unsophisticated use in bandshift assays. The technique is nonquantitative, but when combined with the ability to synthesize oligonucleotides with single base differences, it becomes an immensely potent assay for exploring the DNA–protein interactions that control gene regulation.

References

1. Sambrook, J., Fritsch, E.F. and Maniatis, T. (1989) *Molecular Cloning: A Laboratory Manual*, 2nd Edn. Cold Spring Harbor Laboratory Press, Cold Spring Harbor, pp. 6.36–6.84.
2. Brown, T.A. (1991) *Molecular Biology LabFax*. BIOS Scientific Publishers, Oxford, pp. 261–264.
3. Scott, V., Clark, A.R. and Docherty, K. (1994) in *Methods in Molecular Biology*, Vol. 31, Protocols for Gene Analysis, (A.J. Harwood, ed.). Humana Press Inc., Totowa, pp. 339–347.
4. *Appendix to AT Biochem Handbook* (1994) Protocols and Bibliography for SSCPs and DCSPs.
5. Ainsworth, P.J. and Rodenhiser, D.I. (1994) in *Methods in Molecular Biology*, Vol. 31, Protocols for Gene Analysis, (A.J. Harwood, ed.). Humana Press Inc., Totowa, pp. 205–210.
6. Li, H., Cui, X. and Arnheim, N. (1990) Direct electrophoretic detection of the allelic state of single DNA molecules in human sperm by using the polymerase chain reaction. *Proc. Natl Acad. Sci. USA*, **87**, 4580–4584.
7. Orita, M., Iwahana, H., Kanazawa, H., Hayshi, K., and Sekiya, T. (1989) Detection of polymorphisms of human DNA by gel electrophoresis as single-strand conformation polymorphisms. *Proc. Natl Acad. Sci. USA*, **86**, 2766–2770.
8. Ganguly, A., Rock, M.J. and Prockop, D.J. (1993) Conformation gel electrophoresis for rapid detection of single-base differences in double-stranded PCR products and DNA fragments: evidence for solvent-induced bends in DNA heteroduplexes. *Proc. Natl Acad. Sci. USA*, **90**, 10325–10329.
9. Dumaz, N., Drougard, C., Sarasin, A. and Daya-Grosjean, L. (1993) Specific UV-induced mutation spectrum in the p53 gene of skin tumours from DNA-repair-deficient xeroderma pigmentosum patients. *Proc. Natl Acad. Sci. USA*, **90**, 10529–10533.
10. Watson, J.D., Gilman, M., Witkowski, J. and Zoller, M. (1992) *Recombinant DNA*, 2nd Edn. Scientific American Books, New York. Chap. 18, pp. 335–367.
11. Watson, J.D., Gilman, M., Witkowski, J. and Zoller, M. (1992) *Recombinant DNA*, 2nd Edn. Scientific American Books, New York, pp. 153–174.
12. Abe, M. and Kufe, D. (1993) Characterisation of cis acting elements regulating the transcription of the human DF3 breast carcinoma associated antigen (*MUCI*) gene. *Proc. Natl Acad. Sci. USA*, **90,** 282–286.

11 Denaturing Polyacrylamide Gel Electrophoresis

Applications

Single-stranded DNA molecules.

- DNA sequencing: Research application 11.2.
- DNA footprinting: Research application 11.3.

Single-stranded mRNA molecules.

- RNase protection assays: Research application 11.4.
- Nuclease S1 protection assays: Research application 11.5.
- Primer extension assays: Research application 11.6.

Protocols

Molecular Cloning: A Laboratory Manual. Chap. 13, DNA sequencing pp. 13.2–13.104; Chap. 7, RNase protection assays, pp. 7.71–7.78; S1 nuclease protection assays, pp. 7.58–7.66; Primer extension assays, pp. 7.79–7.83 [1].

Molecular Biology LabFax: Chap. 8, Electrophoresis, pp. 261–264 [2].

Protocols and Catalog from Ambion Inc. (suppliers of reagents for RNase Protection assays). [3].

Methods in Molecular Biology, Vol. 31, S1 mapping using single stranded DNA probes, pp. 299–305 [4].

Methods in Molecular Biology, Vol. 31, DNA footprinting with Rh(phi) $_2$bpy^{3+}, pp. 331–337 [5].

11.1 Overview

Denaturing polyacrylamide gels find a large number of applications in molecular biology and a number of the key techniques are reviewed in this chapter.

- DNA sequencing.
- DNA footprinting.
- RNase protection assays.
- Nuclease S1 protection assays.
- Primer extension assays.

Each of these applications culminate in reactions analyzed on denaturing polyacrylamide gels. However there is only so much that can be gained from knowing how to make and run a particular sort of gel. What is perhaps more important is to gain an understanding of the situations when a particular technique is called for. The same argument is frequently used about statistical tests. The skill is not in being able to carry out a particular calculation; the key is to know which test to apply to which sort of data. The selection of molecular biological applications is in many ways analogous.

Denaturing gels are required for the applications in this chapter because in each case it is a single strand of nucleic acid that is labeled. Generally the size of the single-stranded fragment needs to be measured with some accuracy. As the formation of secondary structure will interfere with separation strictly according to molecular length, denaturing gels are used to prevent any duplex structures from forming. Denaturing polyacrylamide gels are used in preference to denaturing agarose gels because polyacrylamide gels have a smaller pore matrix than agarose gels and offer much higher resolution. Denaturing polyacrylamide gels are prepared with a high concentration of urea to maintain all nucleic acids in a single-stranded form. Sometimes even this fails to remove all base-pairing interactions and a 'sequencing-gel compression' is observed. Solutions to the problem of gel compression are discussed in section 11.2.2. (See Chapter 3 for a discussion of denaturants for nucleic acids.)

11.2 DNA sequencing

11.2.1 The principles of DNA sequencing

There are two distinct methodologies for the determination of DNA sequences which take their names from their inventors. The

Maxam–Gilbert method uses base-specific chemical cleavages to fragment existing DNA molecules [6]. The Sanger method uses base-specific dideoxynucleotides to terminate primer extension reaction randomly [7]. Each will be outlined briefly below. The Sanger method has been adopted as the technique of choice now that oligonucleotide primers and purified polymerase enzymes are widely available. Nowadays, no sequencing project would elect to employ the Maxam–Gilbert method. However, these techniques are used within DNA footprinting studies (see Section 11.3) and occasionally the Maxam–Gilbert technique is used to clarify the sequence in a particularly stubborn region.

Whichever method is adopted, the final result is the generation of a population of single-stranded DNA fragments, where one terminus is constant and the opposite end is randomly terminated or fragmented by a base-specific reagent. The fundamental basis of all DNA sequence determination is the ability of a denaturing polyacrylamide gel to resolve single-stranded molecules to unit nucleotide precision. Practically, this is achievable for molecules up to 1000 bases in length, although no more than about 500 can be resolved at any one time. It is sobering to reflect on the fact that the flagship project to sequence the entire human genome of 3000 Mb, which employs state-of-the-art enzymology, robotics, lasers and computers, depends utterly and completely on the relatively primitive technology of the denaturing polyacrylamide gel. Although there is a fair amount of dexterity involved, each gel can be prepared for no more than a few dollars and could potentially reveal for the first time the product of many thousands of years of human evolution. Although the sheer weight of DNA sequencing information issuing from the Human Genome Project can have a powerfully anesthetic effect, there is undoubtedly something very exciting in being the first individual in the whole human race to see a small section of the human genome as a pattern of autoradiographic bands from your own sequencing gel.

Notwithstanding the above, there are situations where the faithful migration of DNA molecules strictly according to the number of bases in their length breaks down. It is absolutely essential that these occasions are recognized and remedial actions are employed. These gel electrophoresis artifacts are known as 'compressions'. The background to the occurrence of sequencing gel compressions and the steps taken to avoid them will be discussed below. An example of a sequence compression that arose in some of my own work is described as a research application.

Maxam–Gilbert sequencing. This requires a DNA molecule labeled at a single end. The label can be at either the 3' or the 5' end. The 5' labels are generally added using γ-^{32}P ATP and the enzyme polynucleotide kinase (see Section 5.4). Since this treatment will leave a double-stranded molecule labeled at both ends, this molecule will have to be cut with an appropriate restriction enzyme into unequal sizes and the fragments resolved on an agarose or polyacrylamide gel.

The 3' labels can be added by fill-in reactions with α-^{32}P dNTPs and the Klenow fragment of DNA polymerase I. Such molecules will also become labeled at both ends if the molecule was initially released from a vector as a restriction fragment with recessed 3' termini. However, it is possible to use different restriction enzymes to yield either a blunt or an overhanging 3' end, which cannot be used as a template by a DNA polymerase I fill-in reaction. It is also possible to use a selection of dNTPs that label one 3' end but not the other. Of course, a fragment that has been labeled at both ends can be cut asymmetrically and purified from an agarose or polyacrylamide gel. There is no limit to the size of fragment that can be used for Maxam–Gilbert sequencing, although sequence information can only be obtained within 400 bases of the labeled end. There is also no need to begin the reaction with single-stranded DNA, just as long as the molecule is labeled exclusively at one terminus. Further details of end labeling can be found in sections 5.4 and 11.3.

Five chemical cleavage reactions are usually employed in Maxam–Gilbert sequencing. This is because some of the reactions cannot discriminate uniquely between one base and another. The end-labeled material is divided into five aliquots and subjected to the following incubations:

G   Methylation with dimethyl sulfate at pH 8.0;
A + G   Piperidine formate treatment at pH 2.0;
G + T   Hydrazine treatment;
C   Hydrazine treatment at 1.5 M NaCl;
A > C   1.2 M NaOH at 90°C.

Following these reactions the modified DNA samples are incubated in 1 M piperidine at 90°C. This treatment results in the breakage of the phosphate–sugar backbone at the site of one of the base-specific modifications. The concentration and timing of each reaction is critical. It is essential that, on average, only one base is cut in each molecule. In this way a random population of fragments is created whose different fragment lengths correspond to the site of an individual base. Too much reagent and the fragments are too small. Too little reagent and the fragments are too large. Having completed the fragmentation of the end-labeled molecule, the reaction products are separated on a denaturing polyacrylamide gel. Maxam–Gilbert gels can be less straightforward to read than Sanger gels, since at any one length bands occur in more than one lane and a judgment has to be made concerning the relative intensities of the A > C lane.

Sanger sequencing. In contrast to Maxam–Gilbert methodology, Sanger sequencing is centered on the enzymatic copying of the template strand. Therefore it has the potential to work on smaller amounts of material, if steps are taken in the final stages to focus on a single strand. Sanger sequencing can be bolted-on to a PCR strategy for template amplification. This is called direct sequencing [8]. Direct sequencing eliminates the requirement for cloning a DNA or RNA molecule before it is sequenced. A full description of direct sequencing and other PCR amplification methods can be found in ref. 9. Whereas the Maxam–Gilbert technique uses chemical cleavage, the Sanger technique uses base-specific dideoxynucleotides to terminate DNA polymerase randomly and thus mark the position of specific bases in the template. Standard nucleotides in DNA are 2' deoxy on the ribose sugar, but retain the 3' OH group to which the next base is joined (see *Figure 1.2*). Dideoxynucleotides have no 2' or 3' OH groups, so the chain is broken upon incorporation of a dideoxy NTP. (RNA bases have OH groups at the 2' and 3' positions). In Sanger sequencing, the provision of a fixed end to the population of differentially sized fragments comes from the specific oligonucleotide primer, which is bound to the template at a unique location. End labeling of oligonucleotide primers can be employed, but label is more generally incorporated in the form of dNTPs in the polymerase extension reaction. Fluorescently tagged primers and ddNTP terminators are used successfully in automated sequencing (see Section 5.5 and *Figure 5.4* for additional information on dideoxynucleotide sequencing).

Sequencing compressions. In an ideal DNA Sanger sequencing gel, bands would appear in only one of the G, A, T or C lanes at any molecular length. Each band will be equally spaced from the next band and each will be of equal intensity. If all these parameters were adhered to, it would be a simple matter indeed to read the sequence from the gel moving from one lane to another. The ideal image in *Figure 11.1a* illustrates the reading process. However, nature is not always so obliging. Occasionally a region of a gel will give an image like that in *Figure 11.1b*. There is a large gap where no base seems to have caused a termination. Just below the gap there are several bands that are running at the same position. The sequence at this location in the template has been 'compressed' into such a small region of the gel that it is impossible to determine the true sequence. This phenomenon is known as *sequencing compression*. It is of course absolutely essential that this ambiguity is resolved. An extra base or a missing base may lead to the mistaken examination of a large

(a)

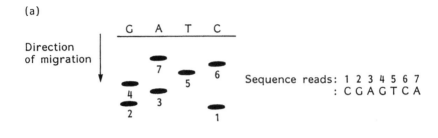

Sequence reads: 1 2 3 4 5 6 7
: C G A G T C A

(b)

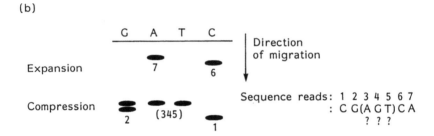

Sequence reads: 1 2 3 4 5 6 7
: C G(A G T)C A
? ? ?

FIGURE 11.1: *The principle of DNA sequencing using dideoxynucleotide termination (Sanger sequencing). Examples of sequencing gel patterns. (a) A small section of an ideal dideoxynucleotide DNA sequencing determination after separation of the single-stranded products of polymerase extension on a denaturing polyacrylamide gel. (b) Sequencing gel compression artifact: secondary structure within single-stranded DNA molecules causes compressed banding patterns in the gel. Sequence expansions are usually observed directly above sequence compressions.*

section of sequence in the wrong reading frame. There are two ways in which a sequencing gel artifact can arise.

(1) Through effects of the template on the behavior of the DNA polymerase during primer extension. Polymerase pausing can lead to artificial termination, which appears as bands across all four lanes. Polymerase pauses are not strictly compressions but they may give rise to similar patterns on the gel.

(2) By vestiges of secondary structure that lead to altered mobilities of single-stranded fragments during denaturing polyacrylamide gel electrophoresis. These are true sequencing compressions. The difference between a polymerase pause and a compression is that a compression is accompanied by a sequence expansion, a gap where there are no bands in the gel above the compression. Polymerase pauses are not preceded by expansions.

It will come as no surprise that polymerase pausing and compressions are more commonplace in GC-rich DNA or in regions where there are large numbers of repeat sequences. Repeated sequences, in particular inverted repeats, have the potential to fold up into secondary structures that may occasionally be strong enough to resist the passage of the polymerase enzyme and the denaturing properties of 6–8 M urea which is present in standard sequencing gels. New generations of DNA polymerases, such as those developed for PCR applications that operate at elevated temperatures, can largely eliminate the first source of sequencing artifact, namely pausing during polymerase extension. At higher temperatures template secondary structure is effectively melted. In contrast, alterations to the performance of the polymerase reaction can have no influence on the occurrence of sequence compressions. These occur due to the conditions that exist **within the polyacrylamide gel**. Compressions occur at the 3' ends of a fragment where there is potential for secondary structure to result in the molecule folding back upon itself. It is thought that this gives rise to a slightly more compact structure, and this permits these fragments to migrate a little faster in the gel, causing the compression artifact. The expansion gap is there because these fragments have migrated a little faster than they should. When the polymerase has passed this region, the folded-back structure is no longer stable and the fragments return to their normal mobilities. Three strategies can be used to overcome sequencing compressions whose origins lie in the process of nucleic acid gel electrophoresis.

(1) *Alterations in gel temperature.* If a denaturing polyacrylamide gel could be maintained at a sufficient temperature, small regions of secondary structure at the 3' ends of the polymerase extension product could be prevented from forming. Unfortunately, at these temperatures urea and polyacrylamide become unstable. Gels are optimally run at 40–50°C, but further elevations are not recommended. Often attaining 50°C uniformly across the gel is not without difficulty. Adhesive liquid crystal thermometers are available to provide an easy means for monitoring gel temperature. The source of thermal energy is Joule heating (see Section 2.1). DNA sequencing gels are run at 1500–2000 V for a 40–50 cm gel, but depending on the design these may need to be pre-run for several hours if they are to attain temperatures of 40–50°C before loading. In some types of apparatus one of the glass plates is attached to a large buffer reservoir that acts as a heat sink. This chamber can be filled with pre-warmed buffer and thus the optimal gel temperature can be attained more rapidly.

(2) *Base analogs.* A more effective strategy, if compressions are still present once attention has been paid to running gels at the optimum temperature, is to replace dGTP with analogs that cannot form as efficient base pairs with C residues. This substantially weakens the potential for secondary structures that

will resist the denaturant power of 8 M urea and eliminates sequencing compression. Two analogs are in common use: dITP (deoxyinosine triphosphate) and 7-deaza-GTP. The structures of these compounds result in weaker base pairs with C residues. DNA polymerases have a lower activity with these unnatural analogs, so the efficiency of the sequencing reactions are often reduced. Nevertheless, by always running a standard dGTP reaction in parallel with a dITP or 7-deaza-GTP reaction the true sequence can be determined. An example of the use of dITP to overcome a sequence compression is described in the research application below.

(3) *Formamide.* An alternative approach to the use of dGTP-based analogs is to make the poylacrylamide more denaturing. This can be achieved by the addition of formamide to the standard urea/polyacrylamide mix (see Chapter 3 for a discussion of nucleic acid denaturants). This procedure effectively eliminates all sequencing compressions. However, the gel solution is more viscous and difficult to pour when it contains formamide. The gel requires a higher voltage and also takes about twice as long to run.

11.2.2   Research application. The use of a dGTP base analog to overcome a sequencing gel compression artifact

Background. An example of a sequence compression is illustrated from previous work carried out in my laboratory. The project centered around the construction of a number of site-directed mutants in a *lacZ* reporter gene encoding β-galactosidase. The wild-type sequence read 5'...CCAGTTCT...3' (see *Figure 11.2A*). We needed to make a mutant 5'...CCAGATCT...3', substituting an A for a T (details of the methods can be found in ref. 10). The autoradiograms of the sequencing gels suggested that this mutant had been constructed, but whenever an A was introduced into the sequence the pattern of bands from the TCT bases following it were perturbed.   Examination of *Figure 11.2B* shows that the first T and the C bands are compressed together and there is a characteristic expansion gap in the banding pattern above. It is difficult to be sure that TCT is still the sequence. An extra C or T may have been inserted, or C and T might have been inverted. (Site-directed mutagenesis methods can throw up all kinds of unexpected surprises, so it is absolutely essential that the DNA sequence is determined with complete confidence.)

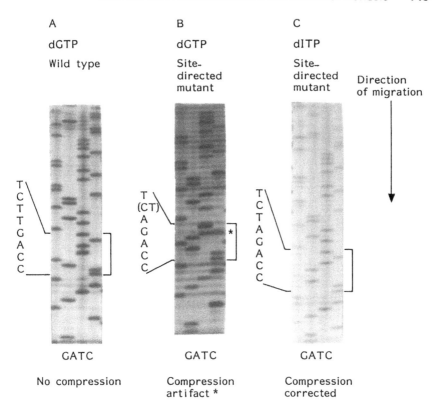

FIGURE 11.2: *The use of a dGTP base analog to overcome a sequencing gel compression artifact.*

Procedure

(1) Sanger method DNA sequencing was performed on double-stranded plasmid DNA using a synthetic oligonucleotide to initiate a primer extension reaction at approximately 35 bases from the site-directed mutation.

(2) Sequencing was performed using a standard kit from Amersham-USB containing the polymerase enzyme Sequenase 2.0. Separate reactions were established using dGTP and dITP and the ddNTP termination products were run in adjacent tracks on a 7 M urea 6%T polyacrylamide gel.

(3) Radiolabel was provided by α-^{35}S thio-dATP, which was present in the primer extension reaction (see Section 5.4 for a description of radioactive nucleotides). Following electrophoresis on a 35-cm gel at 2000 V for 2.5 h the gel was washed and fixed to remove urea, dried and subjected to autoradiography.

Interpretation. Figure 11.2C shows the result of running the primer extension reactions in the presence of dITP in place of dGTP. Note that the dITP tracks are much weaker than the dGTP tracks. This is because the polymerase extends much less efficiently with dITP than with dGTP. However, it is clear that the ambiguous co-migration of the TCT bands that followed the insertion of A in the sequence CCAGATCT is eliminated and the sequence reads as expected. It is interesting to note that the insertion of A created the potential for a repeat structure when CCAGTTCT was altered to CCAGATCT. This may have allowed the following foldback to occur at the 3' end:

5' . . . AGA
   3' TCT

If this was able to form in the sequencing gel, it may have led to the slightly faster migration of the fragments that had this structure at their 3' end. Substituting all the Gs with inosines destabilized the structure and allowed an unambiguous assessment of the site-directed mutation.

11.3 DNA footprinting

11.3.1 The principle of DNA footprinting

Footprinting is the highly descriptive name given to a set of techniques that can be used to find out to which sites proteins bind when they interact with nucleic acids. The basis of the footprinting phenomenon is the protection afforded the nucleic acid from the indiscriminate cleavage of enzymatic or chemical agents. For example, regulatory proteins that bind to upstream promoter regions can be 'footprinted' to reveal their DNA-binding sites. The DNA bears the 'footprint' of the proteins, since the cleaving agent cannot attack the DNA where it is shielded by protein. The principle of DNA footprinting is illustrated in *Figure 11.3*. The essential point to grasp is that the footprint is revealed on a denaturing polyacrylamide gel **as a region where there are no bands**. Related techniques exist in the study of RNA–protein interactions called 'heelprinting' [11] and 'toeprinting' [12]. These methods probe the heel or the toe of a protein footprint by using primer extension reactions that run into the protein from one direction (3' toeprint) or the other (5' heelprint).

All footprinting techniques employ strand-cutting agents that are largely insensitive to DNA sequence. If just the correct amount of cleaving agent is used, then a ladder of bands will be generated as

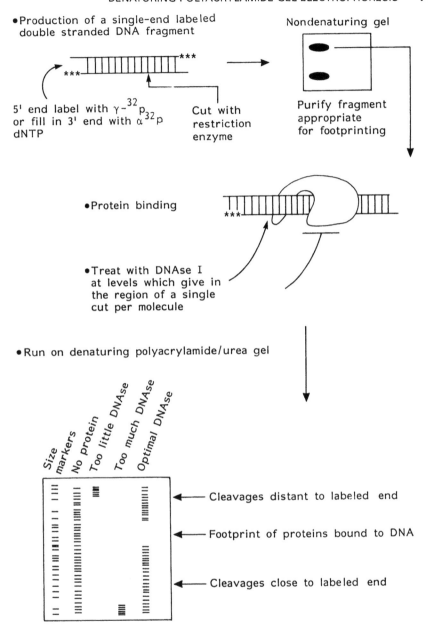

FIGURE 11.3: *The principles of DNA footprinting.*

infrequent cuts are made to the DNA strand at random positions. The endonuclease DNase I is commonly employed to footprint proteins on to DNA. Finer resolutions are obtained by using chemical agents that are smaller and can probe regions which may be inaccessible to a large protein. DNase I is 37 kDa in size. Chemical agents that have been used include agents which generate free radicals [4]. These are generally required when small molecules such as nucleic acid-binding drugs are footprinted. Both enzymatic and chemical cleaving agents have to be titrated. If too much cleaving agent is used, all the strands will be cut to a minimum length and an even ladder of fragments of different sizes will not be produced.

The interpretation of a footprinting experiment is simplified considerably by employing end-labeled DNA. In this way a single cut produces one labeled single-stranded molecule and one unlabeled molecule (see *Figure 11.3*). The unlabeled strand is invisible on the autoradiogram of the denaturing polyacrylamide gel, whereas the labeled strand marks a position in the fragment where the cutting agent was able to gain unhindered access to the phosphodiester backbone. Of course most proteins, and in particular regulatory proteins, recognize and bind to double-stranded DNA structures. The proper substrate for a DNA footprinting assay is therefore a double-stranded molecule labeled at only one end. Such a structure can be generated by labeling a double-stranded fragment at both ends and using a restriction enzyme to cut the molecule in two. The correct fragment can then be purified from a nondenaturing gel. Since the fragments upon which footprinting is generally performed are of the order of a few hundred base pairs, this is often accomplished on nondenaturing polyacrylamide gels (see Chapter 10), which offer greater resolution at low molecular weight than agarose gels (see Chapter 7). Alternative procedures using combinations of gel purification followed by end-specific labeling are equally applicable.

In order to map the protected bases on both strands of a double-stranded molecule, fragments that contain end-labeled strands from both the sense (the same as the mRNA) and the antisense strand must be employed. All techniques that employ gel electrophoresis require suitable size markers to provide a positive control for correct functioning of the gel and to enable experimental fragments to be measured accurately. For DNA footprinting, Maxam–Gilbert sequencing reactions are often used [6]. The Maxam–Gilbert method of DNA sequencing employs the same end-labeled DNA molecules that are used for footprinting reactions. In Maxam–Gilbert sequencing, DNA is subjected to base-specific cleavage reactions, as opposed to the more widely used Sanger method which uses dideoxynucleotides to introduce random terminations during a primer extension reaction [7].

11.3.2 Research application. DNA footprinting

Background. As an example of the application of DNA footprinting, let us return to the paper which illustrated the application of bandshift assays in Chapter 10, Abe and Kufe (1993) 'Characterisation of cis acting elements regulating the transcription of the human DF3 breast carcinoma associated antigen (*MUCI*) gene' [13]. To recap, the aim of the study was to characterize the transcriptional regulation of the *MUCI* gene, which encodes the DF3 glycoprotein at high levels in human breast cancer cells. The complete genomic sequence has been determined for more than 1600 bases upstream from the transcription start site. The study describes reporter gene experiments to map important areas of the promoter region. Having focused on a length of sequence that appeared important for transcriptional activity in MCF-7 breast carcinoma cells, attention was turned to discover where exactly the regulatory proteins might bind within this region. This was achieved using gel retardation or bandshift assays and DNA footprinting. (Gel retardation assays are covered in Chapter 10 using data from the same research paper.) *Figure 11.4* is an autoradiograph of a denaturing polyacrylamide gel used to map protein-binding sites by DNA footprinting.

Procedure

(1) The length of DNA sequence to be footprinted was designated f(–618/–410). The negative numbers indicate the fragment between 618 and 410 bases upstream from the transcriptional start site, which is denoted +1 by convention. This fragment of DNA was cut from a plasmid vector using two different restriction enzymes. DNA polymerase I and α-^{32}P dATP and dTTP were used to fill-in the overhanging 5' ends of an *Eco*RI site:

5' G3'   5' GAATT 3'
3' CTTAA 5'   →   3' CTTAA 5'

(2) The other end of the fragment, which was cut with the second enzyme, did not contain 5' overhanging T or A to act as a template for polymerase in the presence of dATP and dTTP and so would not have been labeled. The double-stranded DF3 promoter fragment was purified from the vector fragment (which would also be end-labeled) on a nondenaturing polyacrylamide gel.

(3) Nuclear proteins were extracted from MCF-7 breast cancer cells and different amounts were incubated with a fixed quantity of the end-labeled, f(–618/–410) double-stranded DNA.

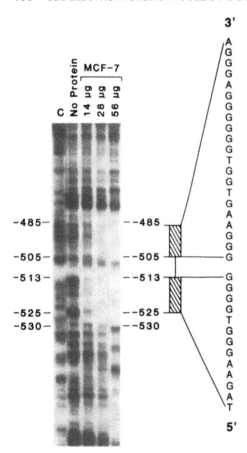

FIGURE 11.4: *DNase 1 analysis of protein binding to f(–618/–410).*
f(–618/–410) was end-labeled (1 x 10⁴ cpm/0.1 ng) and incubated with the
indicated amounts of MCF-7 nuclear proteins. The samples were
subjected to DNase 1 digestion and analyzed in 12% polyacrylamide gels.
C represents analysis of cytosines as determined by Maxam–Gilbert
sequencing. The hatched boxes reflect regions protected from DNase 1
digestion. Reproduced from ref. 13 with permission from the National
Academy of Sciences of the USA.

(4) DNase I was added to the nuclear protein–DNA mixture. A zero
protein reaction was set up as a negative control for footprinting
and as a positive control for DNase I activity.

(5) The products of the DNase I digestion were heated in denaturing
loading buffer and applied to a denaturing polyacrylamide/urea
gel. In these conditions all proteins are stripped off the DNA,
which separates into single strands. Only the strand that carried
the end-specific label will be detected on the autoradiographic
image.

Interpretation. *Figure 11.4* shows the results of the DNA
footprinting analysis of nuclear protein binding to the 5' upstream

region of the gene encoding DF3. The first lane (reading from left to right) is the sequencing size marker. Lane 2 shows the pattern of DNase I cleavage when the fragment was incubated in the absence of nuclear protein. When nuclear proteins are added, lanes 3, 4 and 5, two regions of the gel are cleared of bands indicating protection of the phosphodiester backbone from DNase I. It is not possible to be sure whether this is a single protein or a complex of several proteins. The authors of the paper relate that, as a negative control, these studies were repeated using nuclear extracts from a human cell line which does not express the DF3 protein. No specific DNase I footprints were observed. Further elucidation of the particular proteins that bind to this DNA sequence might take the form of nuclear extract fractionation studies, in order to purify the protein or proteins which leave the observed footprint. An alternative procedure is to carry out a South-Western blot [14]. A South-Western blot is performed as a normal Western blot: a polyacrylamide SDS protein gel is run and the proteins are transferred to a nitrocellulose membrane. However, instead of antibody detection, the membrane is probed with a labeled DNA fragment. Nonspecific DNA binding activities are removed using unlabeled DNA containing unrelated sequences. The ability to remove the labeling of a particular protein with an excess of the unlabeled DNA fragment containing the protein-binding site is an essential control to prove the binding of the labeled fragment is due to the recognition of a specific DNA sequence. This study carried out a South-Western blot revealing that a 45-kDa protein binds to the sequence in question.

Denaturing polyacrylamide gel electrophoresis is essential to a DNA footprinting study to separate the end-labeled fragments according to length, and thus map out positions occupied by the nucleic acid-binding proteins. Nondenaturing polyacrylamide gels are frequently used to purify the single-end-labeled, double-stranded fragments required in the footprinting analysis.

11.4 RNase protection assays

11.4.1 The principle of the RNase protection assay

The principle of the RNase protection assay is shown in *Figure 11.5*. The aim of the assay is to detect the presence of, or to quantify, a specific RNA (usually an mRNA) in a sample of total RNA extracted from a cell or tissue. A 'probe' is generated through *in vitro* transcription. This provides a single-stranded fragment of RNA that can be labeled to high specific activity and can be produced at a defined length (*in vitro* transcription for the production of labeled probes is described in Section 5.4). Briefly, a purified bacteriophage T3, T7 or SP6 RNA polymerase is used to transcribe a DNA template that has been cloned adjacent to the specific bacteriophage promoter. This probe is usually a short fragment, complementary to the sense (mRNA) strand of the target gene. Probes of around 200–500 bp are typical. The probe is mixed, in excess, with the total RNA sample and following a sufficient time for all the mRNA target sequences to be hybridized a mixture of RNase A and RNase T1 enzymes are added. These nucleases cut only at single-stranded regions, so any probe molecules will be destroyed unless they are bound to their target mRNAs. If there is no target mRNA in the sample, all the labeled probe will be degraded. To separate the degraded RNA from the protected probe, the contents of the reaction are applied to a denaturing polyacrylamide gel. Size standards are run on the gel and usually a small sample of the undegraded probe is run as a control. If the probe was radioactively labeled, the gel is dried and an autoradiogram prepared. This can be used to calculate the quantity of mRNA in the sample using densitometry. Alternatively, the film can be used as a guide to cut out the bands from the dried gel; the polyacrylamide can be solubilized with H_2O_2 and the radioactivity measured in a scintillation counter. A more sophisticated method is to employ a phosphorescence imager to calculate the levels of radioactive emission directly in the dried gel.

11.4.2 Research application. RNase protection assay

Background. The paper I have selected to illustrate the principles of the RNAse protection assy describes the isolation of a cDNA from a λ bacteriophage expression library containing cDNAs from the plant species *Arabidopsis thaliana*. Huang *et al.* (1993) 'Characterisation of a gene encoding a Ca2+-ATP-like protein in the plastid envelope' [15]. The library was screened with an antiserum raised to the plastid envelope. This is a double membrane which surrounds the chloroplasts or plastids of higher plants. The cDNA isolated was then used to probe a genomic library. Following the isolation of genomic clones that hybridized to the cDNA, the genomic DNA and cDNA sequences were determined. By comparing the sequences with those already in 'the database', the ever-accumulating computerized database of all known sequences held, shared and collected by EMBL (European Molecular Biology Laboratory) or GenBank, the new *Arabidopsis* gene was found to belong to a family of proteins called P-type ATPases. These proteins are intrinsic membrane pumps that

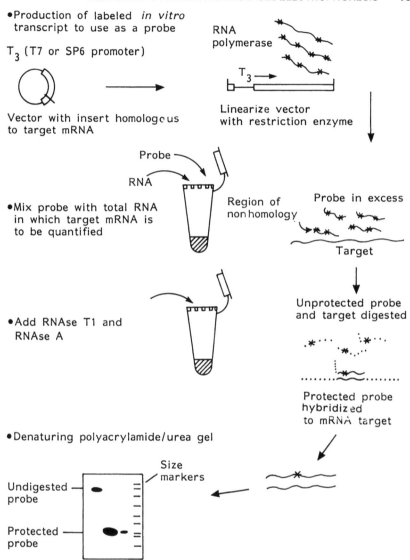

- Production of labeled *in vitro* transcript to use as a probe

T_3 (T7 or SP6 promoter)

RNA polymerase

Vector with insert homologous to target mRNA

Linearize vector with restriction enzyme

Probe

RNA

- Mix probe with total RNA in which target mRNA is to be quantified

Region of nonhomology

Probe in excess

Target

- Add RNAse T1 and RNAse A

Unprotected probe and target digested

Protected probe hybridized to mRNA target

- Denaturing polyacrylamide/urea gel

Size markers

Undigested probe

Protected probe

FIGURE 11.5: *The principles of an RNase assay. The difference in size between the undigested and the protected probe reflects a region of nonhomology between the* in vitro *transcript and the mRNA target.*

maintain a low intracellular Ca^{2+} consuming ATP in the process. The gene described in the paper was designated *PEA1* (plastid envelope ATPase). *Figure 11.6* is an autoradiograph of an RNase protection assay carried out to compare the abundance of the *PEA1* mRNA in leaves with that in roots.

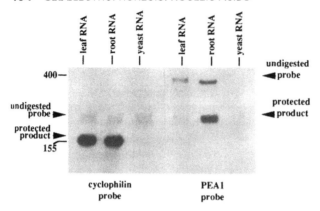

FIGURE 11.6: *Relative transcript levels in roots and leaves determined by RNase protection. Total RNA from* Arabidopsis *leaves,* Arabidopsis *roots, or yeast was probed with ³²P-labeled antisense RNA probes generated either from a cDNA for* Arabidopsis *cytoplasmic cyclophilin as described or from a subcloned 367-bp EcoRI/Sac I fragment from gPEA1. The size of the undigested cyclophilin probe is 204 nucleotides and the expected size of the protected fragment is 160 nucleotides. The size of the undigested PEA1 probe is 375 nucleotides and the expected size of the protected fragment is 213 nucleotides. The molecular sizes of the standards are indicated in nucleotides. Reproduced from ref. 15 with permission from the National Academy of Sciences of the USA.*

Procedure

(1) Total RNA was prepared from *Arabidopsis* leaf and root tissue. Total yeast RNA was prepared as a control, since this would not be expected to contain *PEA1* homologous sequences.

(2) ³²P-labeled *in vitro* transcripts were prepared from a fragment of the *PEA1* gene cloned into a vector where it was adjacent to a bacteriophage promoter. As a control, a fragment of a different *Arabidopsis* gene (cyclophilin) was also prepared as an *in vitro* labeled probe. In previous work cyclophilin had been shown not to vary significantly between leaves and roots and would act as a control for loading, integrity of the RNA and the RNase protection process. The cyclophilin probe acts therefore as the positive control for the process. (Never risk the chance of getting no signal in an experiment without including a positive control. A positive control is a reaction you know should work. Without a positive control you have no idea whether a lack of signal is real or whether it might represent a technical failure of the procedure. The negative control in this experiment was the inclusion of the yeast RNA.)

(3) The probes were mixed with the total RNA preparations and hybridization in solution was allowed to take place. Hybridization

is generally run at an elevated temperature (45°C) in the presence of a denaturant such as formamide to ensure that only completely matched hybrids are allowed to persist.

(4) RNase A, which cleaves after single-stranded U and C, and RNase T1, which cleaves after single-stranded G, are added to the hybridization mixture to destroy all the probe not bound to its homologous target. Note that nonhomology between species or as a result of polymorphisms can lead to shorter lengths of protected probe than expected.

(5) The samples are loaded on a denaturing polyacrylamide/urea gel. Following electrophoresis the gel is dried and autoradiography is performed.

(6) Size standards must be run on the gel, so that the size of the protected fragment can be precisely defined. Often these can be undigested transcripts of known length or sequencing determining reactions of a well-known template.

Interpretation. The autoradiographic image in *Figure 11.6* shows the level of protected probes and indicates directly the quantity of mRNA in the samples. The left-hand panel shows the result when the positive-control cyclophilin probe was used. The levels of mRNA are comparable in both leaf and root, but there is no protection of the probe in yeast RNA as expected. The size of the undigested probe is indicated and there is a trace of undigested probe in each lane. This may reflect residual template DNA in the probe. The protected probe is shorter than the undigested probe because of the presence of vector sequences in the *in vitro* transcript.

In the right-hand panel the level of the protected *PEA1* probe is clearly substantially higher in RNA from root tissue as opposed to the levels in leaves. Again there is no protection following hybridization with yeast RNA. In this case, there is significant protection of the probe at its undigested size. This may indicate the presence of unspliced *PEA1* mRNA, since the *PEA1* probe contains some intron sequences. Quantitation of the radioactivity in the protected bands was carried out using a phosphorescence imaging machine (see Section 5.4). This indicated that there was between seven and nine times as much *PEA1* mRNA in roots as in leaves. Note that it is unsafe to make direct comparisons between the levels of two mRNAs if you cannot be sure that the specific activities of your probes are identical.

Gel electrophoresis was essential for this assay since it permitted the distinction between the probe fragment, corresponding to spliced mRNA with the introns removed, and the probe fragment protected by unspliced RNA. A Northern blot could furnish the same information as the RNase protection assay. However, the protection assay is much more sensitive and straightforward. Indeed, the authors of this paper comment that Northern blotting was tried without success, probably because the *PEA1* mRNA is present at very low levels. The improved sensitivity of the protection assay derives from the fact that a much larger amount of RNA can be hybridized than can be run and blotted from a denaturing agarose gel. Furthermore, hybridization is more efficient when both partners are in solution, as compared to a Northern blot when the target mRNA is immobilized on the filter. RNase protection assays are also more tolerant of degradation in the RNA sample. RNase protection assays are therefore the method of choice for quantitation of an mRNA species but a Northern blot must be used when the full length of the mRNA species needs to be determined. This cannot be achieved by RNase protection, as the probes are typically complementary to only a short, 200–500 bp region of the target mRNA.

11.5 Nuclease S1 protection assays

11.5.1 The principle of nuclease S1 protection assays

The mechanics of a nuclease S1 protection assay, often referred to simply as an 'S1' or as 'S1 mapping', are in many respects similar to the RNase protection assay described above. However, S1 assays use labeled DNA probes in place of RNA probes. Nuclease S1 is a nuclease that is specific for single-stranded DNA, although single-stranded RNA is also hydrolyzed to a lesser extent. Nuclease S1 has traditionally been used to map the 5' and 3' ends of mRNAs and to map the positions of intron–exon boundaries. Although many of these tasks could be accomplished by the newer technique of RNase protection, the S1 technique will be described here since it appears frequently in the literature and it covers the analysis of transcriptional start sites, for which RNase protection could also be used. Most protocols for S1 mapping also call for the purification of the single-stranded DNA probe with the use of a denaturing polyacrylamide gel. This serves to emphasize that gel electrophoresis is often used to purify nucleic acids as well as to analyze them.

The principle of using an S1 assay to map the 5' end of an mRNA is illustrated in *Figure 11.7*. An important and frequently encountered problem in most projects which analyze the structure and function of eukaryotic genes is the recovery of a cDNA with an incomplete 5' end.

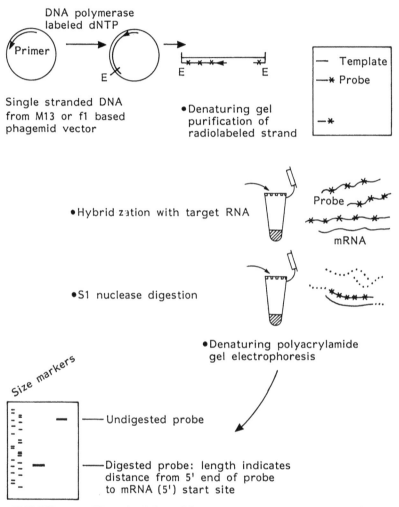

FIGURE 11.7: *The principles of S1 nuclease protection assays for mapping 5′ transcription start sites.*

Reverse transcriptase, the enzyme used to copy mRNA into cDNA, binds to a primer at the 3′ end of the gene (poly-T bound to the mRNA poly-A site). Therefore there is a chance that reverse transcriptase will fall off the template before it reaches the 5′ end of the mRNA, that is the transcriptional start site. (Remember that there may be several hundred bases of untranslated mRNA upstream from the translation start site.) Even when the complete genomic DNA sequence has been determined, after using the cDNA as a probe for a genomic library, it can be difficult to decide exactly where the transcriptional start point is. Knowing where RNA polymerase commences is essential for studies aimed at elucidating the function of promoter and enhancer elements. The critical component of the S1 assay for determining a 5' mRNA terminus is therefore a genomic clone that covers the presumptive transcriptional start site and a portion of the early mRNA coding sequence. The clone is scrutinized so that a probe can be generated by cutting at a downstream restriction site within the coding region, and at an upstream restriction site beyond the point at which transcription is presumed to initiate (see *Figure 11.7*). The probe is hybridized to the RNA sample and nuclease S1 is used to trim back the single-stranded DNA probe where it has not been protected by the 5' end of the message. By running the protected probe on a denaturing polyacrylamide gel, in conjunction with a sequence-determining reaction that offers size markers of single nucleotide spacing, the precise distance between the end of the probe and the 5' end of the message can be defined.

11.5.2 Research application. Nuclease S1 protection assay

Background. The following paper illustrates the application of S1 mapping deals with the characterization of the promoter region of a mouse transcription factor gene, named δ/YY1 for short. Sáfrény and Perry (1993) 'Characterisation of the mouse gene that encodes the δ/YY1/NF-E1/UCRBP transcripton factor' [16]. (The multiple components of the much longer name that appears in the title of the paper represent alternative names given to the same protein by other research groups.) Since there was evidence that the transcription factor itself was regulated, the aim was to examine upstream sequences for evidence of known transcriptional activator binding sites. Clearly the precise point at which transcription begins is crucial to the understanding of the regulation of the δ/YY1 gene by upstream transcription factors. *Figure 11.8* is an autoradiogram of nuclease S1 and Mung bean nuclease mapping of the 5' transcriptional start site. (Mung bean nuclease is a similar enzyme to S1, but because there are sometimes idiosyncratic differences in the specificities of nuclease S1 and mung bean nuclease for different sequences, the two enzymes are often used in parallel and the results compared.)

Procedure

(1) The first step in the S1 assay is to prepare the probe. S1 probes were at one time generated as double-stranded DNA probes of fixed length that were denatured and used in the hope that the

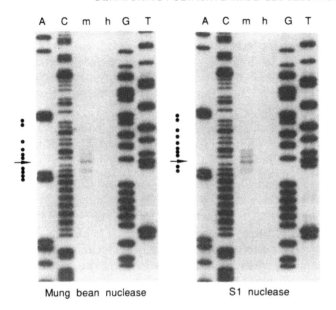

FIGURE 11.8: *Determination of the 5' ends of δ transcripts by nuclease protection. A uniformly labeled 282-nt antisense strand probe (see* Figure 1) *was hybridized to cytoplasmic RNA from S194 mouse (m) or HeLa (h) cells, digested with either mung bean or S1 nuclease, and analyzed by electrophoresis on a 6% polyacrylamide/urea gel together with* ^{32}*P-labeled sequence ladders of the same probe. The arrows and dots indicate the major and minor initiation sites, respectively. Reproduced from ref. 16 with permission from the National Academy of Sciences of the USA.*

greater stability of RNA–DNA heteroduplexes compared to DNA–DNA heteroduplexes would drive the reaction in favor of hybridization to the target mRNA. Single-strand probes are used nowadays. This improves the efficiency of hybridization, since there is no tendency for the probe to anneal to itself. Single-stranded DNA probes can be prepared by primer extension from either double-stranded or single-stranded templates. In the paper under consideration here, a uniformly labeled probe was generated by using Klenow enzyme (a derivative of *E. coli* DNA polymerase I) to extend from a synthetic oligonucleotide primer annealed to a single-stranded phagemid vector, in the presence of α-^{32}P dCTP. The resulting double-stranded molecule was cut with appropriate restriction enzymes, and the labeled single strand of 282 bases in length was purified on a denaturing polyacrylamide/urea gel (this gel purification step is not usually required for RNase protection).

(2) Total RNA was prepared from a mouse plasmacytoma cell line. RNA from the human cell line HeLa was used as a negative

control. The labeled probe was added and hybrids were allowed to form in solution during overnight incubation.

(3) Mung bean nuclease or nuclease S1 was added to digest the unprotected DNA probe.

(4) The digested reactions were then loaded on to denaturing polyacrylamide/urea gels.

(5) DNA sequence ladders of known templates were loaded as size markers (see Section 5.5 and 11.1 for a description of the dideoxynucleotide methods of DNA sequence determination).

(6) Following electrophoresis, the gels were dried and subjected to autoradiography.

Interpretation. *Figure 11.8* depicts the analysis of the 5' transcriptional start site of the δ/YY1 gene using Mung bean nuclease and nuclease S1. As referred to previously, this is sometimes done to guard against misleading idiosyncratic variations in the nuclease specificity of the two enzymes. Here both nucleases give rise to virtually identical patterns. In the left-hand lanes marked A and C, the dideoxynucleotide A and C sequencing reactions are loaded. The template for sequencing was in this case the same as that used to generate the probe. This is clearly a region rich in C because there are many more terminations in the dideoxy C lane than in the dideoxy A reaction. The right-hand two lanes show G and T dideoxynucleotide terminations. This region is clearly rich in G as well as C. Areas of GC-rich sequence are often found at the 5' ends of genes. These areas frequently contain so called 'CpG islands' (CpG means C followed by G in the same strand). CpG is generally rare in mammalian genomes, so gene-rich regions or islands are often recognized by an abundance of CpG [17].

In the central two lanes of both the S1 and mung bean gels, the protected fragment of the labeled DNA probe is shown following hybridization to mouse (m) or human (h) total RNA. The fragment that appears in the mouse lane represents the distance between the 5' end of the mRNA and the (5') end of the probe, in this case 180 bases. If the mRNA was longer at the 5' end, this fragment would be positioned further up the gel. There is on close inspection some heterogeneity in the size of the protected fragment. Since this is sometimes observed for specific promoters but not others, it is believed that this represents flexibility in the transcriptional machinery rather than in the nuclease activity of S1 or mung bean nuclease. There is no protected product observed in the reaction where human RNA was hybridized to the probe. Although the sequences from the two species may be highly homologous (95% identity), single-base differences between probe and template are sufficient to provide mismatched, single-stranded regions that act as cleavage sites for nuclease S1 and mung bean nuclease. Therefore the human reaction acts as a negative control against any kind of self-protection of the probe.

Gel electrophoresis was essential for this study as the single-stranded probe was first purified from contaminating double-stranded DNA by using denaturing polyacrylamide gel electrophoresis. Denaturing gel electrophoresis was then used a second time, to provide high-resolution determination of the precise size of the protected fragment. Only with the degree of resolution provided by denaturing polyacrylamide gels could the transcription start site be accurately mapped.

11.6 Primer extension assays

11.6.1 The principle of the primer extension assay

Primer extension is a widespread concept in molecular biology. All DNA polymerase and reverse transcriptase enzymes require the free 3' OH at the end of a short section of an RNA or DNA strand (a primer) base paired to the template before polymerization can begin. (In contrast, RNA polymerases can initiate template copying in the absence of any primer, if a specific promoter sequence is present.) Primer extension is thus at the heart of DNA squencing, cDNA synthesis, PCR and a number of other techniques. However, the term 'primer extension assay' has a special meaning. In this context, a primer extension assay is generally taken to imply a specific use of template-directed copying to determine the mRNA transcriptional start site.

The principle of the primer extension assay is illustrated in *Figure 11.9*. An oligonucleotide DNA primer is bound to a preparation containing mRNA and a reverse transcriptase enzyme is used to extend the primer until it reaches the end of the template. This is the 5' end of the message and represents the transcriptional start site. Essentially, very similar information is gathered to that from an S1 protection assay. However, in contrast to an S1 assay, a primer extension can be carried out without access to genomic clones. All that is required is sufficient sequence information to be able to synthesize an upstream-facing oligonucleotide primer. The distance to the transcription start site can then be determined.

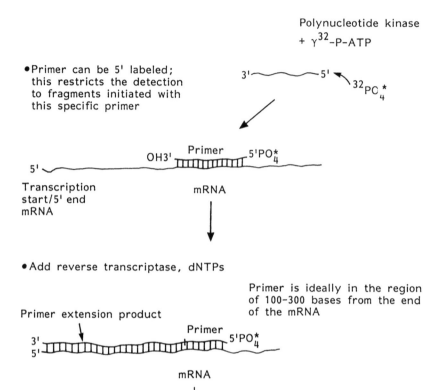

• Primer can be 5' labeled;
  this restricts the detection
  to fragments initiated with
  this specific primer

Polynucleotide kinase
+ γ^{32}-P-ATP

Transcription
start/5' end
mRNA

mRNA

• Add reverse transcriptase, dNTPs

Primer is ideally in the region
of 100–300 bases from the end
of the mRNA

Primer extension product

mRNA

• Run on denaturing polyacrylamide gel

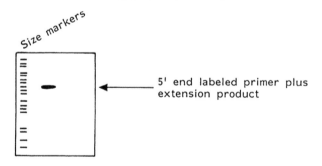

5' end labeled primer plus
extension product

FIGURE 11.9: *The principles of a primer extension assay.*

11.6.2 Research application. Primer extension assays

Background. An example of a research application of the primer extension assay can be found in: Enssle *et al.*, (1993) 'Determination of mRNA fate by different RNA polymerase II promoters' [18]. The study described in this paper was aimed at elucidating why, when alternative RNA polymerase promoters were used to drive expression of the same transcript in mammalian cells, the stability of transcripts containing single-base mutations in the coding region was differentially affected. In other words, when gene A and gene A^{mut} were expressed from promoter 1, the level of transcript A^{mut} is substantially less than that of gene A. When gene A and gene A^{mut} were expressed from promoter 2, both genes generate equal levels of transcripts. In an effort to establish the molecular basis for this effect, primer extension assays were employed to check whether changing the promoter altered the selection of transcriptional start sites. If the start sites were altered, this could explain the effect. *Figure 11.10* is an autoradiogram of a primer extension experiment. The products of the reaction were analyzed on denaturing polyacrylamide gels.

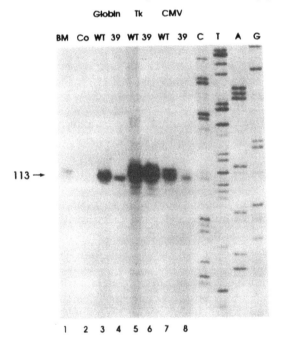

FIGURE 11.10: *Determination of transcription start sites. Primer-extension analysis was performed with RNA from HeLa cells transfected with constructs 1 and 3 (globin, lanes 3 and 4), 7 and 8 (Tk, lanes 5 and 6), or 9 and 10 (CMV, lanes 7 and 8) and of RNA from nontransfected control cells (Co, lane 2) and human bone marrow (BM, lane 1). To allow the visualization of the β-globin and the Tk transcripts without overexposing the CMV signal, we used 10-fold less CMV- than β-globin- or Tk-promoted RNA for this experiment. The DNA sequencing reaction products (lanes C, T, A, and G) were used as size markers. Lanes: WT, normal β-globin gene sequence; 39, codon-39 C → T nonsense mutation. Reproduced from ref. 18 with permission from the National Academy of Sciences of the USA.*

Procedure

(1) Total RNA was prepared from a human cell line called HeLa, which had been transfected with vectors expressing wild-type (wt) or mutant (39) β-globin genes from three alternative promoters: β-globin, thymidine kinase (Tk) and cytomegalovirus (CMV).

(2) A 20-mer oligonucleotide primer, complementary with bases +94 to +113 in the β-globin gene, was labeled at the 5' end using polynucleotide kinase and γ-^{32}P ATP.

(3) The labeled primer was mixed with the samples of total RNA. Moloney leukemia virus reverse transcriptase was added in the presence of dNTPs.

(4) The products of the reactions were loaded on to denaturing polyacrylamide gels. A dideoxy DNA sequencing reaction was loaded as a source of size markers spaced at single nucleotide increments.

(5) Following electrophoresis the gel was dried and an autoradiogram was prepared.

Interpretation. *Figure 11.10* shows the results of primer extension reactions carried out on β-globin genes encoding either wild-type (wt) or mutant (39) alleles. Each was cloned adjacent to one of three promoters: β-globin, Tk or CMV. These vectors were transfected into HeLa cells and the total RNA was extracted and subjected to primer extension using a primer that binds in the β-globin coding sequence. As a positive control, RNA was also extracted from human bone marrow (BM, lane 1). This is expected to be rich in cells expressing β-globin mRNA. RNA from untransfected HeLa cells was subjected to the primer extension reaction to provide a negative control (Co, lane 2). Primer extension gave the expected product of 113 bases in length when carried out on RNA from bone marrow, but there was no primer extension product from untransfected HeLa cells. When the wt and 39 globin alleles were expressed from the globin promoter (lanes 3 and 4) and the CMV promoter (lanes 7 and 8), the transcript with the C→T mutation at codon 39 (base 117) is present in substantially reduced amounts, but the length of the primer extension product remains constant at 113 bases in length. When the Tk promoter is used, the major primer extension product remains at 113, but now the wt and 39 transcripts are expressed at equal levels. The effect is not caused by the total levels of expression, as the CMV promoter is stronger than the Tk promoter: 10-fold less CMV-promoted RNA was used compared to Tk or globin promoted RNA as the signal from the CMV primer extension was so strong.

Gel electrophoresis was essential for this experiment because it allowed the site of transcription initiation to be determined with primer extension to single-base resolution, by running the products on high-resolution denaturing polyacrylamide gels. Whilst there are many other methods that could be employed to measure the level of expression, such as RNase protection, Northern blotting or dot-blot hybridization, primer extension permitted a semi-quantitative assessment of expression, at the same time as the determination of the transcription start site.

References

1. Sambrook, J., Fritsch, E.F. and Maniatis, T. (1989) *Molecular Cloning: A Laboratory Manual,* 2nd Edn. Cold Spring Harbor Laboratory Press, Cold Spring Harbor.
2. Brown, T.A. (1991) *Molecular Biology LabFax.* BIOS Scientific Publishers, Oxford, pp. 261–264.
3. Ambion Inc. Catalog (1994).
4. Viville, S. and Mantovani, R. (1994) in *Methods in Molecular Biology,* Vol. 31, Protocols for Gene Analysis, (A.J. Harwood, ed.). Humana Press Inc., Totowa, pp. 299–305.
5. Klakamp. S.L. and Barton, J.K. (1994) in *Methods in Molecular Biology,* Vol. 31, Protocols for Gene Analysis, (A.J. Harwood, ed.). Humana Press Inc., Totowa, pp. 331–337.
6. Maxam, A.M. and Gilbert, W. (1977) A new method of sequencing DNA. *Proc. Natl Acad. Sci. USA,* **74,** 560–564.
7. Sanger, F., Nicklen, S. and Coulson, A.R. (1977) DNA sequencing with chain-terminating inhibitors. *Proc. Natl Acad. Sci. USA,* **74,** 5463–5467.
8. Phear, G.A. and Harwood, J. (1994) in *Methods in Molecular Biology,* Vol. 31, Protocols for Gene Analysis, (A.J. Harwood, ed.). Humana Press Inc., Totowa, pp. 247–256.
9. Newton, C.R. and Graham, A. (1994) *PCR.* BIOS Scientific Publishers, Oxford.
10. Phillips-Jones, M.K., Hill, L.S.J., Atkinson, J. and Martin, R. (1995) Context effects on misreading and suppression at UAG codons in human cells. *Mol. Cell. Biol.,* **15,** pp. 6593–6600.
11. Kim, J-K. and Hollingsworth, M.J. (1992) Localization of *in vivo* ribosome pause sites. *Anal. Biochem.,* **206,** 183–188.
12. Ringquist, S., MacDonald, M., Gibson, T. and Gold, L. (1993) Nature of the ribosomal mRNA track: Analysis of ribosome-binding sites containing different sequences and secondary structure. *Biochemistry,* **32,** 10254–10262.
13. Abe, M. and Kufe, D. (1993) Characterisation of cis acting elements regulating the transcription of the human DF3 breast carcinoma associated antigen (*MUCI*) gene. *Proc. Natl Acad. Sci. USA,* **90,** 282–286.
14. Philippe, J. (1994) in *Methods in Molecular Biology,* Vol. 31, Protocols for Gene Analysis, (A.J. Harwood, ed.). Humana Press Inc., Totowa, pp. 349–361.
15. Huang, L., Berkelman, T., Franklin, A.E. and Hoffman, N.E. (1993) Characterisation of a gene encoding a Ca^{2+}-ATP-like protein in the plastid envelope. *Proc. Natl Acad. Sci. USA,* **90,** 10066–10070.
16. Sáfrány, G. and Perry, R.P. (1993) Characterisation of the mouse gene that encodes the δ/YY1/NF-E1/UCRBP transcription factor. *Proc. Natl Acad. Sci. USA,* **90,** 5559–5563.
17. Larsen, F., Gundersen, G., Lopez, F. and Prydz, H. (1992) CpG islands as gene markers in the human genome. *Genomics,* **13,** 1095–1107.
18. Enssle, J., Kugler, W., Hentze, M.W. and Kulozik, A.E. (1993) Determination of mRNA fate by different RNA polymerase II promoters. *Proc. Natl Acad. Sci. USA,* **90,** 10091–10095.

Appendix A

Glossary

Allele: the correct term when referring to alternative versions of a gene which exists in a population or are created in the laboratory.

Antisense: a description of a length of sequence complementary to the strand which serves as mRNA. The strand identical to the mRNA is known as 'sense'.

Bacteriophage: a virus which infects bacteria. Many different bacteriophages have been adapted for use as vectors for cloning and propagating DNA molecules.

cDNA: a copy of mRNA made by the enzyme reverse transcriptase, cDNA has no introns and none of the control sequences which are present at the 5' ends of genes.

DNAse: enzymes which degrade DNA

Dot blot: a rapid procedure to test whether a particular sequence of mRNA, cDNA or genomic DNA is present in a sample by spotting a small volume on to a nitrocellulose membrane. The presence of the target sequence is revealed by hybridization with a labeled complementary probe.

Downstream: a phrase used to describe features to the 3' side of any particular point, when viewing a fragment of genomic, mRNA or cDNA sequence in the conventional orientation i.e. with the beginning (5') of a gene to the left and the end (3') of a gene to the right. In this orientation the top strand of double-stranded molecule is the sense (mRNA) strand and the lower strand is the antisense (mRNA template) strand.

Exons: blocks of sequence which actually contain the genetic information specifying a gene product such as a protein. Exons are interrupted by introns in most eukaryotic and a few prokaryotic organisms. Introns are removed in a process called splicing.

Expression library: a collection of cDNA inserts in a vector representing the mRNA molecules present in a particular tissue, where the vector has been designed to permit transcription of the cDNA and translation into protein. If means are available to identify production of the protein, for example by a specific antibody, then individual cDNA clones encoding the protein can be isolated. λgt11 is a vector from which an expression library can be created.

λgt11: a vector derived from the bacteriophage λ that infects *E. coli.* cDNA molecules can be inserted in the vector using restriction enzymes. In bacteria infected with λgt11, cDNA can be expressed and screens can be devised to identify the proteins specified by the cDNA.

Genome: the entire complement of nucleic acid containing the genetic information for a particular organism.

Genomic clone: a segment of the genome inserted into a vector for propagation. All control sequences and introns are present.

Introns: blocks of sequence that interrupt the genetic information (exons) which specifies a gene product such as a protein in most eukaryotic and a few prokaryotic organisms. Introns are removed in a process called splicing.

Library: a collection of cDNA or genomic inserts in a plasmid, bacteriophage or phagemid vector. Unlike a library which holds books, cDNA or genomic libraries are **unordered** populations of hundreds of thousands of cloned molecules representing the entire mRNA (cDNA) or DNA (genomic) complement of a particular organism or tissue.

mRNA: messenger RNA. The mature RNA copy of the genetic information contained within a gene which will be decoded by ribosomes into protein. In mRNA the introns have been removed.

Oligo dT: a short length of synthetic DNA used to purify the mRNA fraction of total RNA by binding to 3' poly A tails.

Oligonucleotide: a short section of nucleic acid analogous to the term 'peptide' when referring to proteins. Oligonucleotides can be synthesized rapidly in machines and are indispensable in modern molecular biology as PCR primers, DNA sequencing primers, hybridization probes and sequences for carrying out site directed mutagenesis.

Phagemid: a vector which can be propagated as a plasmid or as a bacteriophage when used in different conditions.

Plaque: an area of lysis on a lawn of cells which have been infected with a virus.

Plasmid: a circular molecule which is able to replicate extra chromosomally. Plasmids are often used as vectors to propagate and express cloned cDNAs.

3' poly A: a tail of A is added to the 3' end of most eukaryotic mRNAs. Poly A tails can extend for hundreds of residues and are involved in controlling the efficiency of translation into protein.

Polymorphism: a description of a characteristic which can exist in a variety of alternative forms in a population.

Reporter gene: a technique to examine the function of a block of sequence for control of gene expression. An easily assayed gene is inserted downstream of the test region so that the response can be conveniently followed.

Restriction enzyme: enzymes which are purified from different species of bacteria and used in molecular biology to cut DNA molecules at specific sequences. In the host they protect from viral infection by recognizing base sequences in the genome of the incoming organism. The host bacterial DNA is rendered resistant by base specific methylation.

Restriction length fragment polymorphism (RFLP): the term used to describe sequence differences between two otherwise identical regions of the genome which result in the creation or loss of a restriction enzyme site, and consequently, a change in the length of DNA molecules when cut with this restriction enzyme.

RNAse: enzymes which degrade RNA.

rRNA: ribosomal RNA. Ribosomes contain a larger proportion of RNA than protein. There are two main rRNA molecules which are often used as size markers for denaturing agarose gels. In mammals these are the 28S and 18S rRNA, and in bacteria the 23S and 16S rRNA.

RT-PCR: Reverse transcriptase-Polymerase Chain Reaction. The procedure in which a PCR amplification is carried out on cDNA prepared by reverse transcription of a total RNA or mRNA sample.

Sense: a description of a length of sequence identical to the strand which serves as mRNA. The complementary strand is known as antisense.

Sequenase 2.0: a genetically engineered DNA polymerase from bacteriophage T7 used extensively for DNA sequencing by Sanger methodology.

Splicing: the removal of intervening sequences 'introns' from the RNA copy of a gene and the joining together of the 'exons' which encode the mRNA. As cDNA is a copy of mRNA, cDNA has no introns.

Total RNA: the RNA in a cell is mostly rRNA and tRNA. mRNA makes up a smaller fraction. mRNA can be purified from eukaryotic cells by adsorbing to oligo dT via the 3' poly A tails.

Upstream: a phrase used to describe features to the 5' side of any particular point, when viewing a fragment of genomic, mRNA or cDNA sequence in the conventional orientation i.e. with the beginning (5') of a gene to the left and the end (3') of a gene to the right. In this orientation the top strand of a double-stranded molecule is the sense (mRNA) strand and the lower strand is the antisense (mRNA template) strand.

Vector: the means by which a sequence is cloned, propagated or transferred between bacterial strains or different organisms. Plasmids, bacteriophages and phagemids are all vectors.

Appendix B

Suppliers

There appears below a list of suppliers of reagents and apparatus for nucleic acid gel electrophoresis and detection. The location of the main company is given along with a Fax number and an internet address if available. A message to the main site will elicit the whereabouts of the nearest branch or local distributor for your laboratory. The inclusion of suppliers in this list should not be taken as an indication that they have superior products to suppliers which are not included.

Apparatus and reagents

BioRad Laboratories, 2000 Alfred Nobel Drive, Hercules, CA 94547, USA. Fax 510 741 1060.

Gibco-BRL Life Technologies Inc., PO Box 6009, Gaithersburg, MD 20884 9980, USA. Fax 301 670 8539.

Pharmacia Biotech, Box 776, S-191 27 Sollentuna, Sweden. Fax 623 00 69.

Sigma Chemical Company, PO Box 14508, St Louis, MO 63178, USA. Fax 314 771 5757. http://www.sigma.sial.com

Stratagene, 11011, North Torrey Pines Road, La Jolla, CA 92037, USA. Fax 619 535 0045. www site planned for 1996

Reagents

Ambion Inc., 2130 Woodward St, Suite 200, Austin, TX 78744, USA. Fax 512 445 7139.

Boehringer Mannheim GmbH, Sandhofer Strasse 116, D-68298 Mannheim, Germany. Fax 621 759 2890.

FMC BioProducts, 191 Thomaston Street, Rockland, ME 04841, USA. Fax 207 594 3491.

Molecular Probes Inc., PO Box 22010, Eugene, OR 97402 0414, USA. Fax 503 344 6504.

New England Biolabs Inc., 32 Tozer Road, Beverly, MA 01915, USA. Fax 508 921 1350. http://www.neb.com

Promega Corporation, 2800 Woods Hollow Road, Madison, WI 53711 5399, USA. Fax 608 277 2516. http://www.promega.com

Reagents, radioisotopes and safety apparatus

Amersham International plc, Amersham Place, Little Chalfont, Bucks HP7 9NA, UK. Fax 1494 542266. www site planned for 1996

Mergers and take-overs

In recent times the following companies have merged with companies listed above.

AT Biochem	now	FMC BioProducts.
Hoefer	now	Pharmacia Biotech.
USB	now	Amersham International plc.

Index